Employee Experience & Engagement with Predictive Analytics

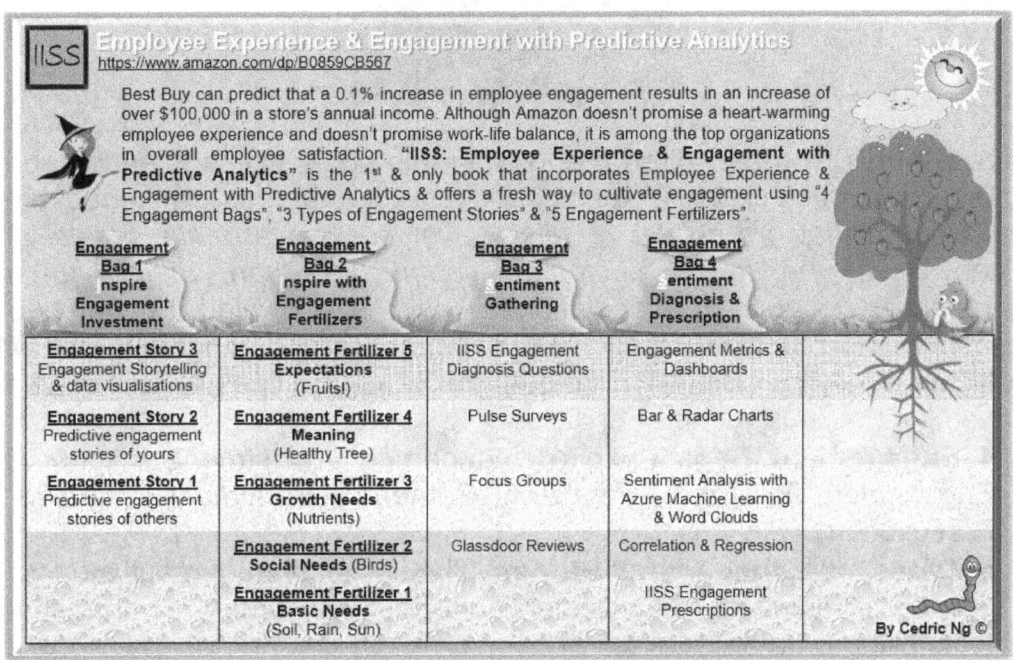

Copyright by Ng Mong Shen 2020

Acknowledgements

Dave Ulrich's words inspired me to write this book, which is endorsed by Dave. Dave once said, *"Expectations shape attitudes. Attitudes drive behaviors. Behaviors deliver results. Lower end restaurants, hotels, and stores have far fewer customer complaints than their high-end counterparts because of expectations"*. – with this in mind, I developed a fresh way to cultivate employee experience and engagement with predictive analytics using 4 Engagement Bags, 3 Types of Engagement Stories, and 5 Engagement Fertilizers:

- **4 Engagement Bags** (Inspire Engagement Investment, Inspire with Engagement Fertilizers, Sentiment Gathering, Sentiment Diagnosis & Prescription).
- **3 Types of Engagement Stories** (Predictive engagement stories of others, Predictive engagement stories of yours, Engagement storytelling & data visualisation).
- **5 Engagement Fertilizers** (Basic Needs, Social Needs, Developmental Needs, Meaning, Expectations).

I would also like to thank my family for their support and understanding.

About the author

The author, Cedric Ng Mong Shen has more than 20 years Global HR experience in various Top Fortune 100 US/European/Asian MNCs, covering Technology, Manufacturing, Oil & Gas, Logistics, and Hospitality Service industries.

Cedric has a Master's in Business Administration (MBA) from University of Strathclyde, and a Bachelor in Economics & Sociology from National University of Singapore.

As a HR thought leader, he has published several HR books sold by Amazon worldwide in markets covering North and South America, Europe, Middle-East, Africa, and Asia-Pacific.

As a Regional HR Trainer and Consultant, Cedric also conducts HR workshops in various countries (Singapore, Malaysia, Thailand, Korea, etc.) on topics covering Predictive HR Analytics, Employee Experience & Engagement, Change Management, Global HR practices, Organizational Design, Total Rewards, Salary Structure & Sales Incentives Design.

Praise for "IISS: Employee Experience & Engagement with Predictive Analytics" book

Exceptional!!!!! This is truly great work. There are some things I really really like:
- Building engagement on the past. So many things I read make it sound like engagement is a completely "new" topic (e.g. some of the work on experience). You have positioned the evolution of the idea very nicely.
- Linking to key business outcomes ... I really like your work on Employee Engagement and Customer Engagement and other business outcomes.
- Using statistics and analytics to identify more subtle insights.
- Offering ways to visualize and use the engagement work.

Again, marvelous work. Thanks for taking the time to write it.
By Dave Ulrich
Speaker, Author, Professor, United States

Nice employee engagement framework! I had a chance to pre-read Cedric Ng Mong Shen take on engagement. I think it offers a nice framework for understanding how employees interact with an organization. Very good read!
By Eric Torigian
VP Global HR/Chief HR Officer at Akebono Brake Corporation, United States

Very Informative! As a HR professional, I find this book very useful in how predictive analytics can be used to measure employee engagement. As more people use tools such as surveys to measure employee engagement, this is one way to really maximize the knowledge gained and to show how statistical methods can really be a power tool in gaining insights.
By Tomeka Hill-Thomas, PhD, Analytics Leadership Council Board Member at DMA, Associate Director of Advance People Analytics at EY, United States

A very interesting reading! Cedric does a wonderful job revising different theories on engagement theory. A great resource if you're looking for different perspectives to tackle engagement and employee experience for your company.
Sergio Garcia Mora
Argentina's People Analytics Authority, Argentina

Cedric's book is easy to read and very insightful! He decodes the employee experience and engagement journey in great clarity and simplicity. Love the analogy of engagement bags and engagement fertilizers! Highly recommended!
By Christian Neubert,
Co-Founder & Managing Director at Human Edge, Switzerland

The author takes what is a complicated topic and makes it easily understandable! Beyond just the goal of decoding employee experience and engagement, he provides analytical tools to bring science to a human behavioral topic!
By Tony Gomes,
Corporate Vice President, Global HR at LG Electronics, United States

This book gives the reader an overview of employee engagement with a theoretical background and perspectives on the importance of engagement. The relevant analytical tools are explained in a clear and concise manner with practical examples. This book is an essential guide to all HR professionals who wish to link organizational decision making to data and analytics.
By Asiyath Mohamed,
Head of Employee Experience at Bank of Maldives

Very much engaging! Employee experience and engagement with predictive analytics is a fantastic introduction to new adventurous in the employee experience field, and a remarkable guide for us working on EX.
By Victoria Polanco,
Employee Experience Manager at Amazon, Costa Rica

The IISS book explains how run correlations, regression analysis and interpret data. It is a great book that does not scare people with complex analysis of t-tests or Anova, and gives you refreshing ideas on how to improve employee engagement!
By Mark Barrera Simcox,
Director at Evolutek, Mexico

Employee Engagement explained as never before! The author brilliantly explains how to cultivate Employee Engagement in your organization from a scientific and analytic point of view. A step by step and data driven approach that will help your success where other books and authors fail. This book is a must if you are into People Analytics. As an HR Professional, and People Analytics Lecturer myself, I absolutely recommend this.
By Mariano Alonso González,
HR Analytics Author/Lecturer at Hábilon, Spain

As an HR academic researcher and industry practitioner I really appreciated the book "IISS: Employee Experience & Engagement with Predictive Analytics" elaborating on advanced topics of workforce analytics and focusing on employee engagement. Among others it presents a straightforward data-driven methodology including impressive visualizations and easily usable tools such as Microsoft Excel.
By George Panagopoulos,
Technology Transfer Consultant at Praxi Network, Greece

A must have for any HR concerned with employee engagement. This book will give you a broad and full understanding of employee engagement, from theoretical basics to data storytelling and recommendation, with middle steps about data survey and analysis, practical examples with excel, etc., following a driving way using seed growing metaphor to illustrate. A must have for any HR concerned with employee engagement! Easy to read and ready to apply.
By Charles Fournier,
People Analytics, France

A must read for HR managers and data scientists. Companies invest a big deal of time and resources making employees happy but as this book explains it doesn't mean they will work hard for the organization. The IISS book integrates Predictive Analytics with Employee Motivation theory and offers a refreshing way to cultivate employee engagement with 4 Engagement Bags!
By Guillermo de la Hoz,
People Director, Spain

Good to read for any HR Professional! 100% recommended! I love how the writer link Engagement with data analytics. Nowadays, providing data about soft HR areas is more than crucial.
By Albert Balague Acebrón
Regional HR Manager at DHL Supply Chain, Spain

Employee experience and predictive analytics are important topics and here they are explained in a very understandable form. The predictive analytics techniques taught in the IISS book will make it easy to convince the management to invest more in employee engagement programs!
By Annika Heino
Human Resources at Geological Survey of Finland

A great Step by Step scientific data driven approach. Superb explanation on employee engagement in the organization. A great reference for HR analytics / HRIS.
By Usman Bashir Ahmed
Total Rewards Head, Saudi Arabia

Great book on the building blocks of employee experience analytics! The author provides a comprehensive overview of employee experience concepts, best practices and motivation theories using examples from various organizations. Next, the reader is gives step-by-step guidance for analysis via MS Excel. As usual, Cedric's books are highly readable, introduces the subject well and give practical advice on how to conduct meaningful analysis with simple Excel tools.
By Ayelet Artzi
Organizational Psychologist and People Analytics Expert at ActiView, Israel

I have been reading the clear path laid out by the author starting with the investment, to the setting up, to the use and finally the diagnosis of engagement with predictive analytics. The explanation of how to use excel for the correlation and regression analysis is easy follow and will not scare away practitioners. The examples are clean and clear. In particular, making sense of the data and how-to diagnosis and interpret the results is worth the read.
By Steven Lum
Adjunct Faculty at NYU School of Professional Studies, United States

As a HR professional, enhancing employee experience and engagement is a key initiative these days. This book gives us comprehensive insights: Theory, Real-world Business Practice, Organizational Strategies to improve employee experience and engagement, how to analyze data from A to Z using Excel. This book helps me navigate in a right direction to scratch from ground for HR data analysis.
By Lynn Jung
Economics Researcher at Indian Business Research Centre, United States

Learn how to boost employee experience with analytics. Simple, yet powerful. This IISS Book is a must read for every HR professional. Highly recommended!
By Praveen Viswam, GPHR
Analytics Consultant, India

Best employee experience and engagement book ever! Highly Recommended.
By Vitthal
Regional HR Head, Agile Airport Services Pvt Ltd, India

The creation of good visual links in the IISS book really helps to highlight the power and value of people analytics.
By Karen Johnson
Human Resources Researcher, United Kingdom

If you'd like to learn about how you can advance your analytics function but with a focus on how to improve employee engagement, then this is a great read. Your Chief HR Officer will appreciate the insights you'll start generating and use it to develop strategic HR policy changes and programs. You'll be a superstar in no time!
By Jared Bester
Compensation Manager at Canada Life, Canada

What I liked about the book is its practical value - even if you are not an expert in HR or analytics, you can use Cedric's detailed guidelines as a blueprint and run your own little people analytics shop in MS Excel. Running regression to predict revenue per employee based on the employee engagement model becomes a walk in the park once you get your data in shape.
By Luka Babic
CEO at Orgnostic, Serbia

Another gem by Cedric! A must for HR professionals! If you are an HR professional and you haven't read one of Cedric's books, you really should. In all his books, Cedric does a wonderful job of defining an HR problem or issue and then showing the reader, through extremely clear and easy instructions, how to use Excel's data tools, from simple scatter point diagrams to regression analysis, to gain insight into the issue. He doesn't bog the reader down in statistical theory or analysis; he simply tells the reader how to interpret the results. If you are an HR professional and want to better understand your employees, but you lack the analytical or statistical skills to understand an HR issue, this is the book for you. Cedric provides clear, step-by-step instructions on how to analyze data regarding these issues, using Excel's data tools, to predict outcomes. Most importantly, he tells you what the results actually means. Conversely, if you have a statistical background but lack the HR context to apply it, to Cedric covers topics such as employee engagement, employee satisfaction, mentoring and finding in work and how data analysis can help predict trends or outcomes. Anyone from college students to seasoned professionals will get something out of this book. I highly recommend it!
By Donald G LeBlanc
Director of Staff at California Air National Guard, United States

Great book! This is a great, easy to read book that explains key metrics that we should all be looking at if we take employee engagement seriously. Definitely pick this book up and give it a read.
By Sabrina Martinez
Workforce Analytics & Reporting Manager, at Memorial Hermann Health System, United States

Whether you are a data person or not, this book shows the perfect interconnectedness between the exact things shown by data and soft part of HR activities. Perfect book to learn a lot of things and start thinking outside the box, and implement many different things based on what you already have in your sources, but you've never thought about them. Recommending!
By Tara Babić,
Compensation and Benefits Analyst at Česká pojišťovna, Serbia

Great Read! This is a really great reference guide for the HR professional. Across the globe, HR has been transitioning to rely more heavily on data for decision making. Turnover data, retention rates, survey data, and more, all give important insights to the ROI of the firms most important resource - People! This book does a fabulous job of outlining areas to track and how to do so. I have found multiple concepts in this book that I intend to put to practice in the segments of Reward and Global Mobility.
By Stephanie Baker,
Compensation at Fransican Health, United States

Making Engagement results useful! Research has repeatedly shown that employee engagement drives organization profit. Cedric has put employee experience and engagement into a great theoretical framework and provides refreshing insights on how to turn engagement results into actions through predictive analytics, a niche skill that is catching on in the evolution of HR.
By Nicholas chew
People Analytics Consultant at MLC Life Insurance, Australia

I especially love the way the engagement fertilizers are explained. Very relatable and easy to understand.
By Temitayo Samson-Grace
HR Business Partner at Rensource Energy, Nigeria

Amazing book! Very useful and helped me to understand in concepts related to engagement, with clear explanations.
By Leonardo Bravo
HR Specialist at Voiter, Brazil

The approach of the 4 engagement bags is really insightful. It helps practitioners to expand their views of how to build the employee journey from ground zero and within the big organizational picture, where often, several silent factors of engagement are underestimated.
By Rosario Sheen
Professor (MBA Program) at Universidad Peruana de Ciencias Aplicadas, Peru

I am a big fan of this book! It is the answer to all the questions that you might have regarding employee engagement and is a must-read for all HR practitioners. The author shares expert insights that provide a wealth of knowledge on how to enhance employee experience and engagement with Predictive analytics. I love the fresh and inspiring perspective that Cedric offers to improve the most strategic HR intervention, linking it to positive business outcomes. I highly recommend this book!
By Rhea Banerjee
HR Director – Global Leadership team at 180Degrees, India

I was happy to read another book from Cedric on EX and EA. It contains practical cases in Excel (using Data analytics tool), which can be a great tutorial for other people analytic working issues. A few models (self-determination theory, Maslow hierarchy of needs, Herzberg's motivation theory etc., describing in the book, will also broaden all the perspectives on Employee Experience & Engagement, helping to see a big picture. Step-by-step, Cedric is guiding on a deep level the understanding and solving issues both on the surface and inside the practice.
By Marianna Simonian
HR Analyst at Yva.ai, Russia

Table of Contents

Table of Contents 10
About this book 13
1.0 Introduction 17
   1.1 Overview of IISS 17
      Engagement Bag 1: Inspire Engagement Investment 18
      Engagement Bag 2: Inspire with Engagement Fertilizers 20
      Engagement Bag 3: Sentiment Gathering 28
      Engagement Bag 4: Sentiment Diagnosis & Prescription 30
   1.2 What is Employee Experience & Engagement? 33
      Employee Satisfaction 34
      Employee Engagement 34
      Employee Experience 34
      Employee Life Cycle 35
      Three types of employees 36
      Employee Net Promoter Score (eNPS) 37
   1.3 Climate vs Culture vs Engagement 39
2.0 Engagement Bag 1: Inspire Engagement Investment 40
   2.1 Engagement Story 1: Predictive engagement stories of others 42
   2.2 Engagement Story 2: Predictive engagement stories of yours 52
      2.2.1 Analytics Maturity Model 53
      2.2.2 Correlation analysis of engagement investment 58
      2.2.3 Regression analysis of engagement investment 66
   2.3 Engagement Story 3: Engagement Storytelling & data visualisation 75
      2.3.1 Data storytelling 76
      2.3.2 Stories & visuals that you can use to engage people 77
      2.3.3 Develop engaging stories 90
      2.3.4 Develop engaging visuals 94
3.0 Engagement Bag 2: Inspire with Engagement Fertilizers 106
   3.1 Engagement & Motivation Theories 108
      3.1.1 Maslow's Hierarchy of Needs 110
      3.1.2 Herzberg's Motivation Theory 117
      3.1.3 Self-Determination Theory 123
      3.1.4 Autonomy, Mastery, Purpose Framework 126
      3.1.5 Engagement MAGIC model 130
   3.2 Engagement Fertilizer 1: Basic Needs 132

- 3.2.1 Compensation 133
- 3.2.2 Benefits 139
- 3.2.3 Work Conditions 147
- 3.3 Engagement Fertilizer 2: Social Needs 150
  - 3.3.1 Quality of friends 152
  - 3.3.2 Help people make friends with others 154
  - 3.3.3 Help people make friends with you 158
  - 3.3.4 Recognition 160
- 3.4 Engagement Fertilizer 3: Growth Needs 162
  - 3.4.1 Teach others 165
  - 3.4.2 Train Managers to Lead 166
  - 3.4.3 Gamify your workplace 167
  - 3.4.4 Mobile learning 168
  - 3.4.5 Motivational talks 169
  - 3.4.6 Orientation & onboarding 169
  - 3.4.7 Job rotation 171
  - 3.4.8 Coaching 174
  - 3.4.9 Mentoring 175
- 3.5 Engagement Fertilizer 4: Meaning 189
  - 3.5.1 Meaning in engaging work 197
  - 3.5.2 Meaning in work that helps others 200
  - 3.5.3 Meaning in work you're good at 202
  - 3.5.4 Meaning with friends at work 204
  - 3.5.5 Meaning in work that fits with your life 206
  - 3.5.6 Meaning in organizational values 207
  - 3.5.7 Meaning in employee value proposition 215
- 3.6 Engagement Fertilizer 5: Expectations 221
  - 3.6.1 Expectancy Theory 223
  - 3.6.2 Equity Theory 225
  - 3.6.3 Expectation Gap 228
  - 3.6.4 Roles of management & employees in aligning expectations 229
  - 3.6.5 How to align expectations? 231
- 4.0 Engagement Bag 3: Sentiment Gathering 236
  - 4.1 Organisational engagement survey 239
    - 4.1.1 Gallup survey questions 241
    - 4.1.2 Aon Hewitt survey questions 243
    - 4.1.3 Officevibe survey questions 246

- 4.1.4 IISS Engagement Diagnosis Questions 250
- 4.2 Pulse surveys 251
- 4.3 Focus groups 252
- 4.4 Glassdoor reviews 253
- 5.0 Engagement Bag 4: Sentiment Diagnosis & Prescription 255
  - 5.1 Steps to analyze engagement surveys 256
  - 5.2 Engagement Metrics & Dashboards 259
    - 5.2.1 Engagement Metrics 259
    - 5.2.2 Engagement Dashboards 267
  - 5.3 Bar Charts 283
    - (i) Year-on-year comparison 283
    - (ii) Business unit vs Market average vs Global average 285
  - 5.4 Radar Charts 287
  - 5.5 Correlation analysis of engagement prescription 292
  - 5.6 Regression analysis of engagement prescription 300
  - 5.7 Sentiment analysis 309
    - 5.7.1 Real-World Impact of Sentiment Analysis 313
    - 5.7.2 Run Sentiment Analysis in Excel with Azure Machine Learning 314
    - 5.7.3 Correlation Example: Determine relationship between "Glassdoor Company Ratings" and "Company Attrition Rate". 325
    - 5.7.4 Regression Example: Predict "Company Attrition Rate" with "Glassdoor Company Ratings" 332
  - 5.8 Word Clouds 339
  - 5.9 IISS Engagement Prescriptions 348
- Annex 1) Publications by Author 349
- Index 350

About this book

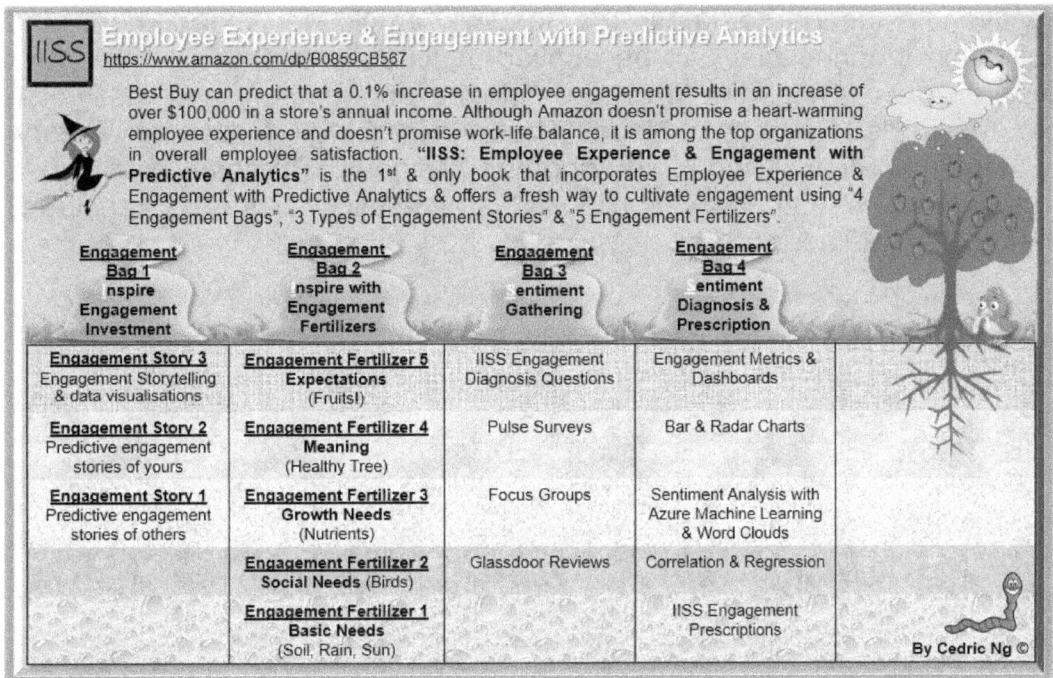

Best Buy can predict that a 0.1% increase in employee engagement results in an increase of over $100,000 in a store's annual income. Although Amazon didn't promise a heart-warming employee experience and didn't promise work-life balance, it is among the top organizations in overall employee satisfaction. **Endorsed by Dave Ulrich, IISS is the 1st & only book that incorporates Employee Experience & Engagement with Predictive Analytics.** It offers a fresh way to cultivate engagement using "4 Engagement Bags", "3 Types of Engagement Stories" & 5 Engagement Fertilizers".:

1) Introduction: Understand the difference between Organisational Climate, Organisational Culture, Employee Engagement, Employee Experience & Employee Satisfaction.

2) Overview of all the leading employee engagement and motivation theories: To be a subject matter expert for employee engagement, HR practitioners needs to know the leading employee engagement and motivation theories that can be applied in different situations. This book gives an overview all the leading employee engagement and motivation (e.g. Maslow's Hierarchy of Needs Theory, Herzberg's Motivation Theory, Self-Determination Theory, Autonomy, Mastery, Purpose framework, Engagement MAGIC Model, Job Characteristics Model, Expectancy Theory, Equity Theory).

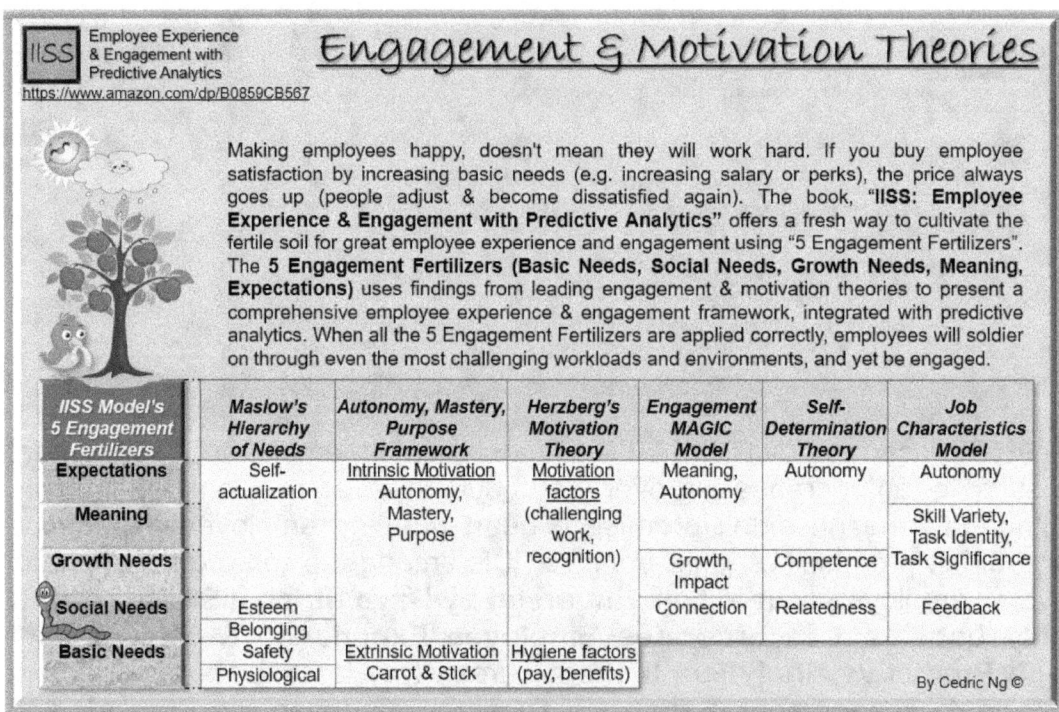

3) Engagement Bag 1: Inspire with Engagement Investment: Despite what businesses say in their mission statements, they don't exist just to "care for our employees' well-being". Ultimately, most businesses care about engagement only if it helps them make money. That's why a significant portion of the IISS model is devoted to show how engagement improves people's performance and business results using predictive analytics and stories. Learn how to inspire Engagement Investment with "3 types of Engagement Stories" (Predictive engagement stories of others, Predictive engagement stories of yours, Engagement storytelling & data visualisation). Learn how to predict the impact of Employee Engagement on Revenue, Customer Satisfaction, Workplace Accidents with Excel using predictive analytics techniques such as Correlation & Regression.

4) Engagement Bag 2: Inspire with Engagement Fertilizers: Making employees happy, doesn't mean they will work hard for you. If you buy employee satisfaction by increasing basic needs (e.g. increasing salary or perks), the price always goes up (people adjust & become dissatisfied again). IISS offers a fresh way to cultivate the fertile soil for great employee experience and engagement using "5 Engagement Fertilizers". The IISS model & it's **5 Engagement Fertilizers** (Basic Needs, Social Needs, Growth Needs, Meaning, Expectations) is aligned with leading engagement & motivation theories, & it uses findings from leading theories & research to present a comprehensive employee experience & engagement framework, integrated with predictive analytics. When all the 5 Engagement Fertilizers are applied correctly, employees will soldier on through even the most challenging workloads and environments, and yet be engaged.

- **Engagement Fertilizer 1: Basic Needs** – Soil, Rain, Sun
- **Engagement Fertilizer 2: Social Needs** – Birds
- **Engagement Fertilizer 3: Growth Needs** – Nutrients
- **Engagement Fertilizer 4: Meaning** – Healthy Tree
- **Engagement Fertilizer 5: Expectations** – Fruits!

5) Engagement Bag 3: Sentiment Gathering: Learn how to use Organisation-wide surveys, Pulse Surveys, Focus Groups, Glassdoor Reviews, IISS Engagement Diagnosis Questions.

6) Engagement Bag 4: Sentiment Diagnosis & Prescription: Learn how to use Engagement Metrics & Dashboards, Bar Charts, Radar Charts, Correlation, Regression, IISS Engagement Prescriptions, Sentiment Analysis with **"Azure Machine Learning"** & Word Clouds.

1.0 Introduction

1.1 Overview of IISS

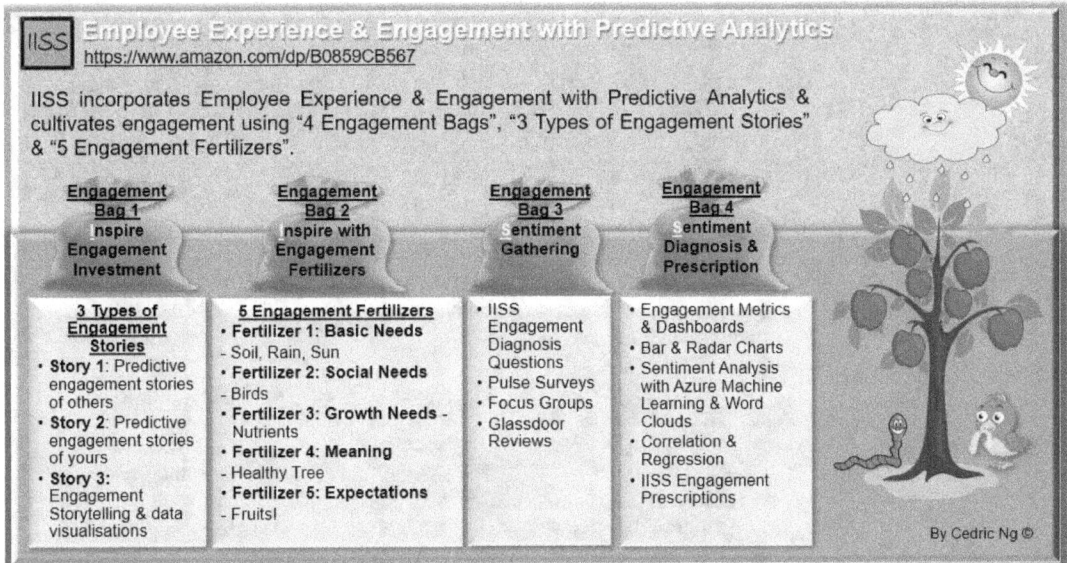

Best Buy can predict that a 0.1% increase in employee engagement results in an increase of over $100,000 in a store's annual income. Although Amazon doesn't promise a heart-warming employee experience and doesn't promise work-life balance, it is among the top organizations in overall employee satisfaction based on data gathered from Glassdoor.com. **IISS is the 1st & only book that incorporates Employee Experience & Engagement with Predictive Analytics.** It offers a fresh way to cultivate engagement using "4 Engagement Bags", "3 Types of Engagement Stories" & 5 Engagement Fertilizers".

- **Engagement Bag 1: Inspire Engagement Investment**
- **Engagement Bag 2: Inspire with Engagement Fertilizers**
- **Engagement Bag 3: Sentiment Gathering**
- **Engagement Bag 4: Sentiment Diagnosis & Prescription**

Engagement Bag 1: Inspire Engagement Investment

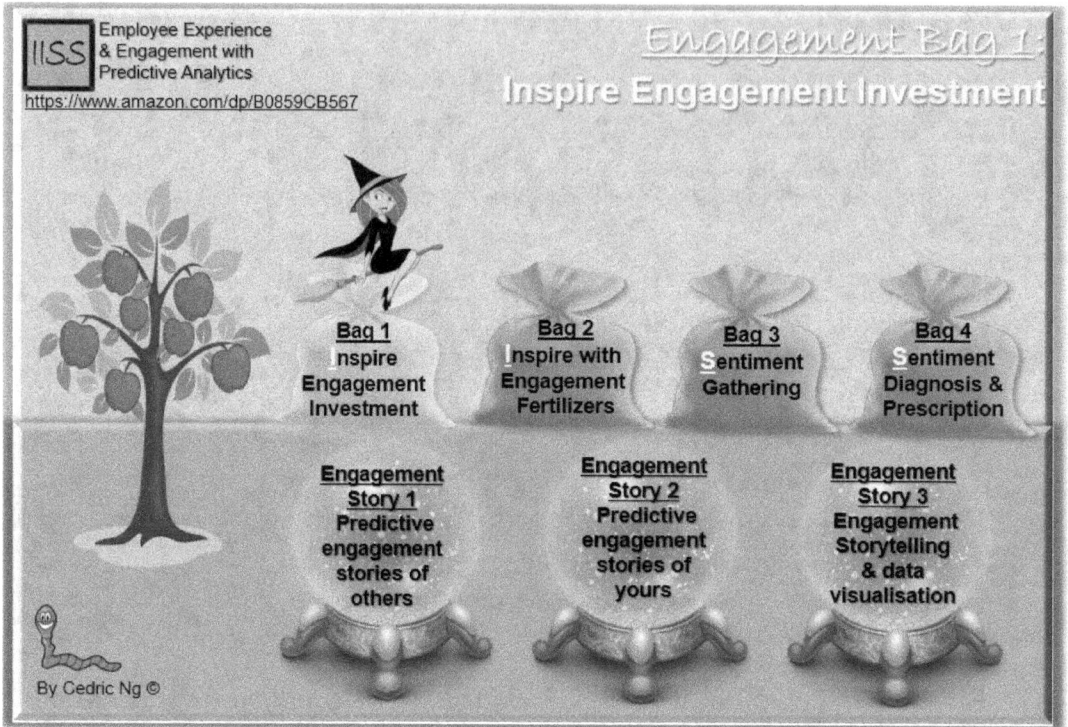

In IISS, **"Inspire Engagement Investment"** is the first engagement bag because the organization and managers need to know how engagement benefits them, before they support it. This section covers how to Inspire Engagement Investment with "3 Types of Engagement Stories":
- **Engagement Story 1) Predictive engagement stories of others**
- **Engagement Story 2) Predictive engagement stories of yours**
- **Engagement Story 3) Engagement storytelling & data visualisation**

Apple farm: To convince organizations to invest in your apple farm, you need to show them how sweet the apples can be, if they invest in fertilizers.

Organization: Similarly, to convince organizations to invest in employee engagement programs, you need to show them how organizational objectives can be met, if they invest in employee engagement programs. Despite what businesses say in their mission statements, they don't exist just to "care for our employees' well-being". Ultimately, most businesses care about engagement only if it helps them make money. That's why a significant portion of this book is devoted to show how engagement improves people's performance and business results.

Apple: To convince people to buy your apples, you need to show them how sweet the apples can be, if they take care of their apple tree.

Managers: Similarly, to convince managers to engage their staff, you need to show them how their team objectives can be met, if they take care of their employees.

Engagement Bag 2: Inspire with Engagement Fertilizers

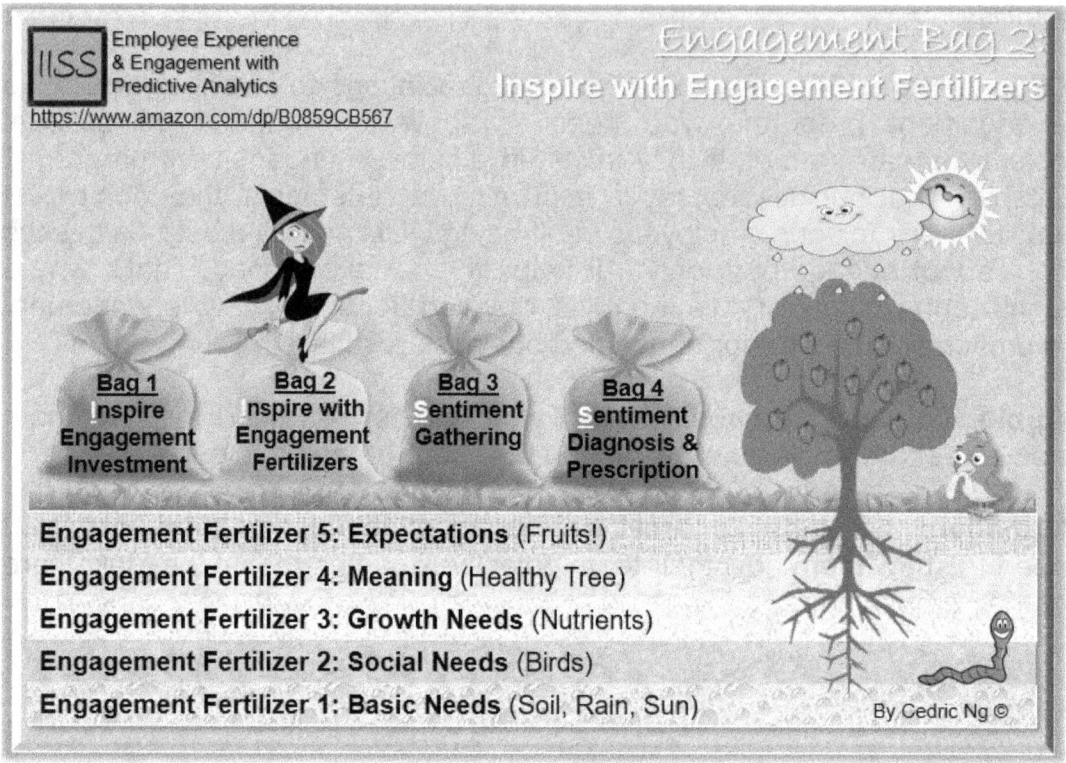

In IISS, "**Inspire with Engagement Fertilizers**" is the second bag, because you can only successfully implement engagement programs, after you get your organization and managers support. Making employees happy, doesn't mean they will work hard for the organization. In IISS, "**5 Engagement Fertilizers**" are used to create the fertile soil for great employee experience and engagement.
- **Engagement Fertilizer 1: Basic Needs** - Soil, Rain, Sun
- **Engagement Fertilizer 2: Social Needs** - Birds
- **Engagement Fertilizer 3: Developmental Needs** - Nutrients
- **Engagement Fertilizer 4: Meaning** - Healthy Tree
- **Engagement Fertilizer 5: Expectations** - Fruits!

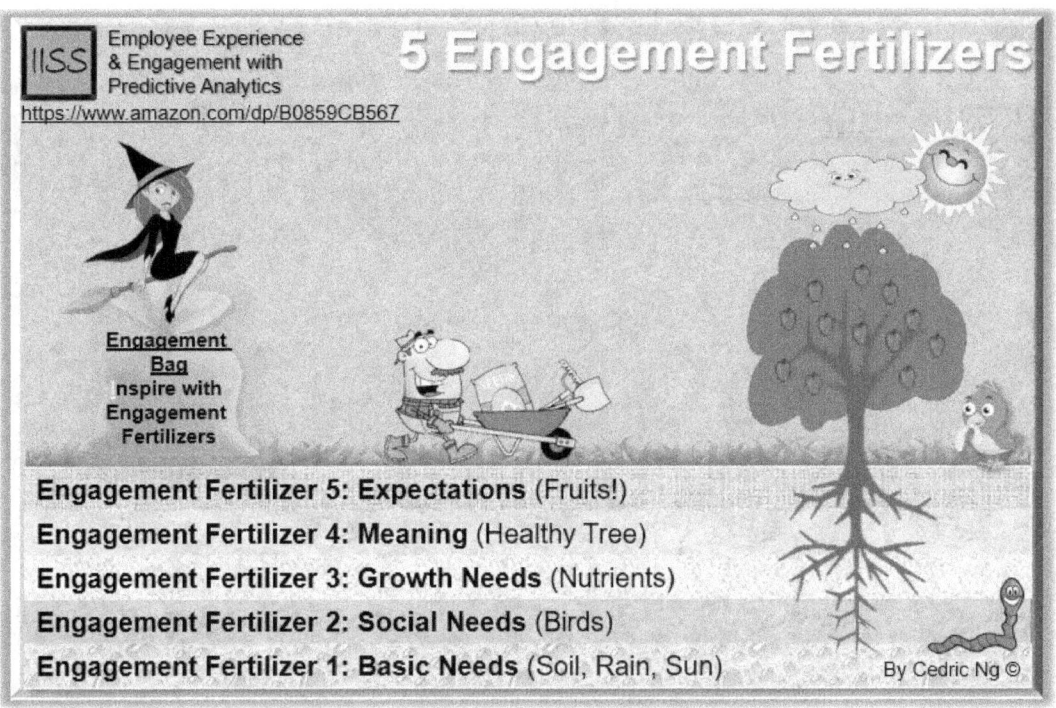

Although the 5 "Engagement Fertilizers" are described as hierarchical, application of the Fertilizers need not be rigid. Without fertilizer 1 (soil, rain, sun), you can't grow satisfaction. But to grow engagement, you need the rest of the fertilizers (right amount of rain, sun, nutrients, and the help of birds).

Engagement Fertilizer 1: Basic Needs (Soil, Rain, Sun)

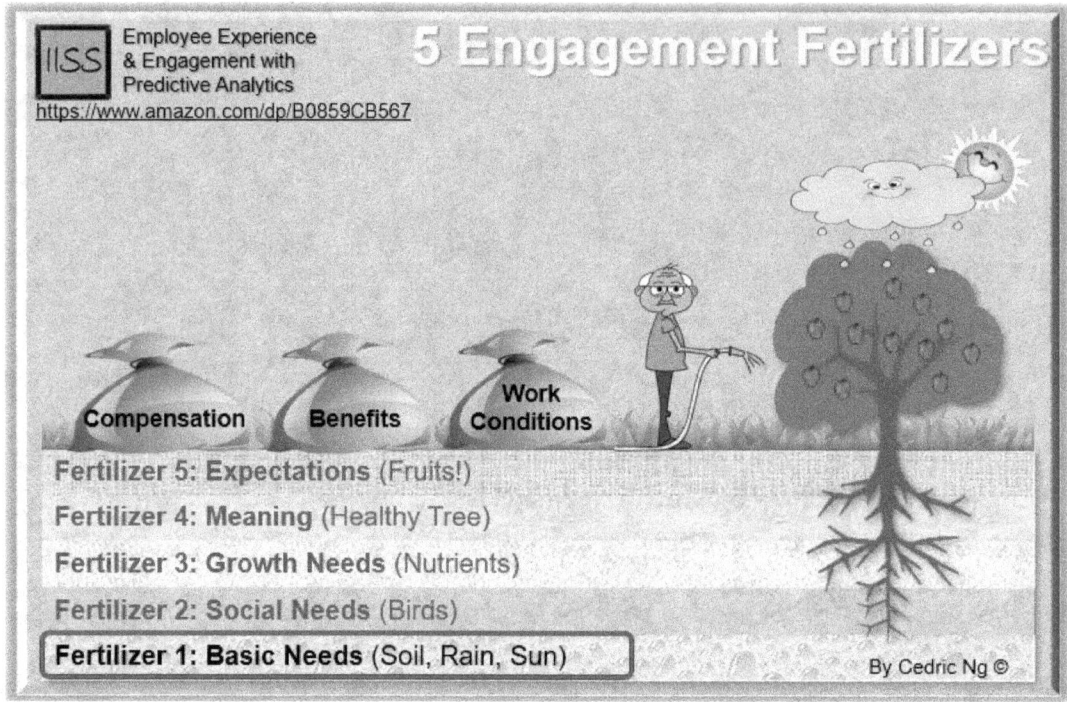

Apple tree: In order for your tree to survive and grow, it needs engagement fertilizer 1 (soil, rain, and sun).

Organisation: Similarly, to have satisfied employees, you need to meet their basic needs for salary, benefits and work conditions. People become dissatisfied when they feel that the basic needs they should be entitled to, is not there, or has been taken away. However, constantly introducing more and better basic needs doesn't increase job satisfaction or performance much because of the "Principle of Adaptation". When people move to a higher income level or higher living standard, they adjust and become dissatisfied again. If you buy employee satisfaction by increasing basic needs and perks, the price always goes up. Money and perks matter - it's hard to be engaged when you feel underpaid or shortchanged. But money isn't why people love their jobs, and although perks are important, they don't engage.

Engagement Fertilizer 2: Social Needs (Birds)

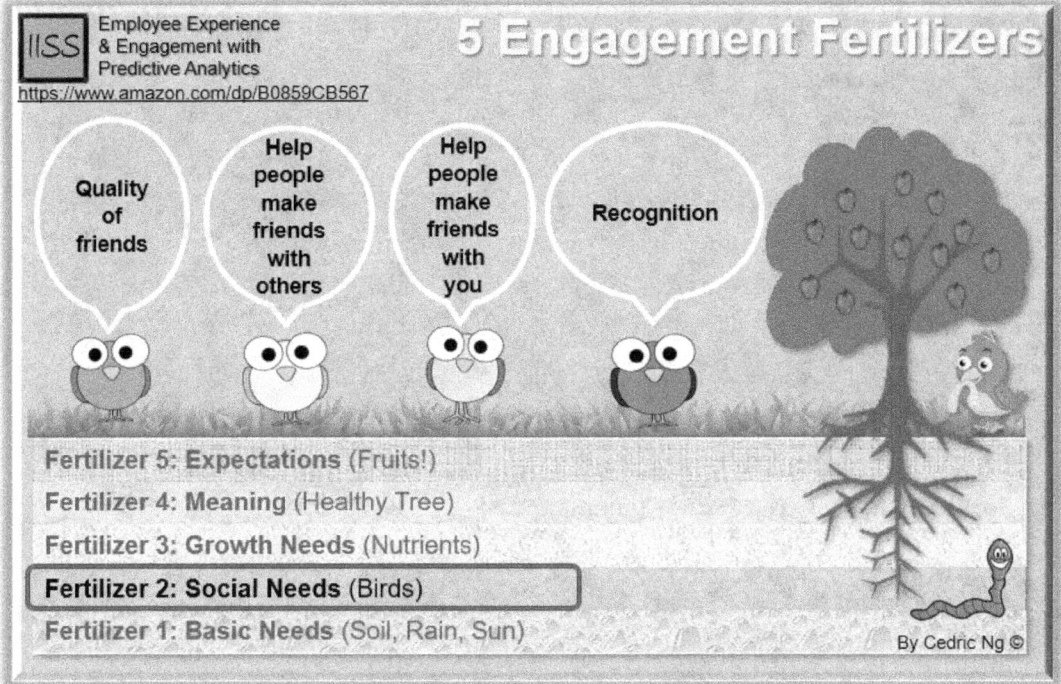

Apple tree: Engagement Fertilizer 2 (Birds) helps to keep your fruit tree healthy by getting rid of the worms that eat the apples.

Organisation: Similarly, people need friends to keep them engaged. Mixing with disgruntled employees can destroy engagement.

Engagement Fertilizer 3: Growth Needs (Nutrients)

Apple tree: Your tree will grow bigger and you'll get bigger apples, if you provide your apple tree with engagement fertilizer 3 (Growth Needs).

Organisation: Similarly, you will get more engaged employees, if you satisfy their growth needs.

Engagement Fertilizer 4: Meaning (Healthy Tree)

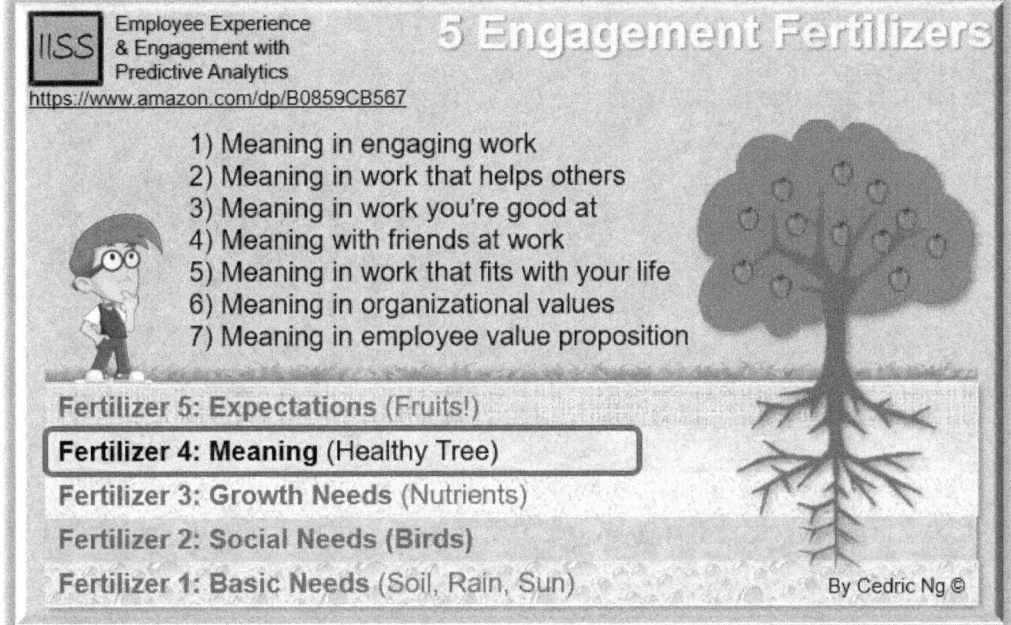

Apple tree: For a Tree to be healthy, you need to provide the right type of fertilizers. Using the wrong type of fertilizers can kill the tree.

Organisation: Similarly, employees working in the "coolest places to work on the planet" can be disengaged if they cannot find meaning in their work. Organizations can till the soil and lay down the fertilizers that allow people to create their own meaning out of mundane work through:
- Meaning in engaging work
- Meaning in work that helps others
- Meaning in work you're good at
- Meaning with friends at work
- Meaning in work that fits with your life
- Meaning in values
- Meaning in employee value proposition

Engagement Fertilizer 5: Expectations (Fruits!)

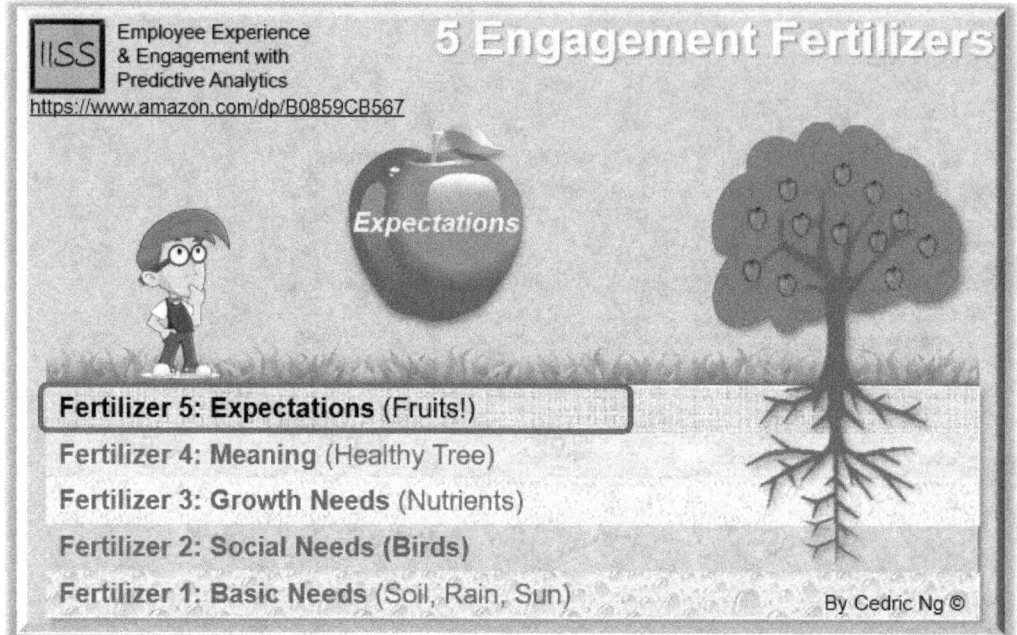

Apple tree: To get big and sweet apples, you need to provide the right amount of sunlight and water (alignment of expectations). Different trees need different amount of sunlight and water (different expectations). Too much or too little sunlight and water can kill the apple tree (misalignment of expectations can kill the apple tree).

Organisation: Similarly, the Organization and Employee have different expectations for Basic Needs, Social Needs, Growth Needs, Meaning, Autonomy, Equity, etc. – Both the Organization and Employee expectations have to be aligned and met for employee experience and engagement to flourish. Employees will soldier on through even the most challenging workloads and environments, if both the Organization and Employee expectations are realistic and aligned.

Engagement and disengagement are more contagious than viruses. You can quarantine a person with a flu virus to prevent it from spreading. But how do you quarantine an employee infected with disengagement? You can't. You either try to engage them or remove them. If you did neither, they will infect their colleagues with negativity. When the 5 Engagement Fertilizers (Basic Needs, Social Needs, Growth Needs, Meaning, Expectations) are applied in the workplace, it builds disengagement immunity, and fuels employee's experience and engagement.

Engagement Bag 3: Sentiment Gathering

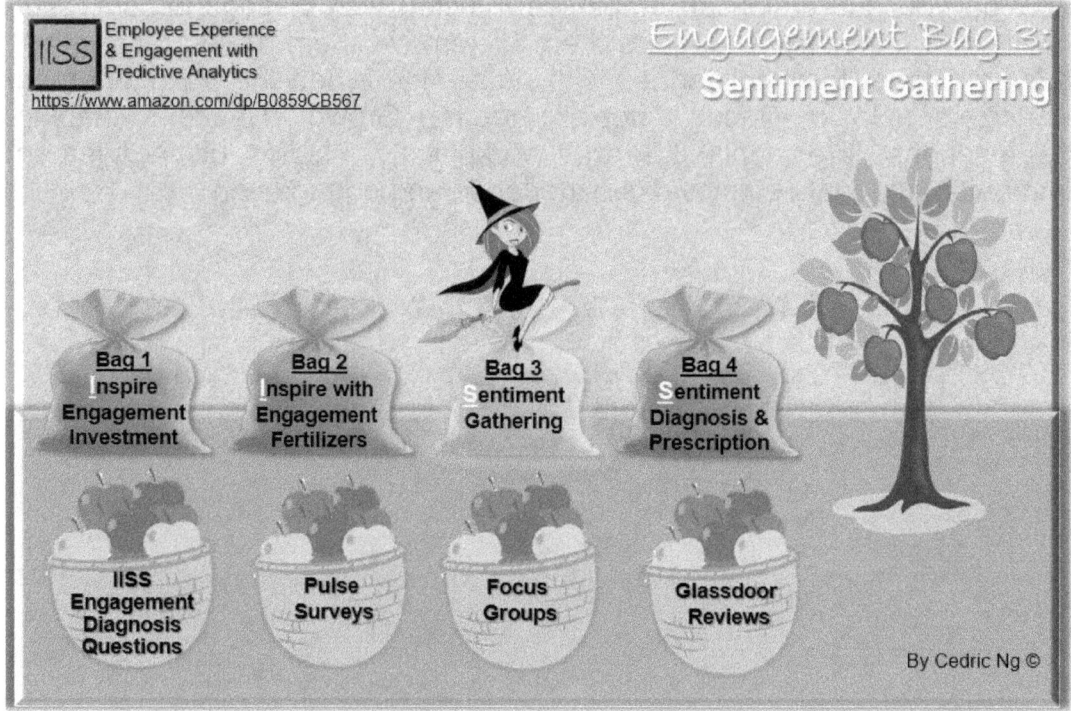

In IISS, "**Sentiment Gathering**" is the third engagement bag because you can start evaluating the effectiveness of the engagement programs after you've launched it. There are various ways to gather employee sentiment. This section covers these approaches to gather employee sentiment:
- IISS Engagement Diagnosis Questions
- Pulse surveys
- Focus groups
- Glassdoor reviews

Apple tree: You need to gather your apples to know if they are sweet or sour. There are various ways to gather apples - you can gather all the apples once a year, or you can gather apples from a section of your apple farm on an adhoc basis, or you can gather just a few apples to sample, or you can check customer feedback on how your apple taste.

Organisation: Similarly, you need to gather employee sentiment to know if your employees are engaged or not. There are various ways to gather employee sentiment - annual organizational wide surveys, adhoc pulse surveys, focus groups, Glassdoor reviews, and IISS Engagement Diagnosis Questions.

Engagement Bag 4: Sentiment Diagnosis & Prescription

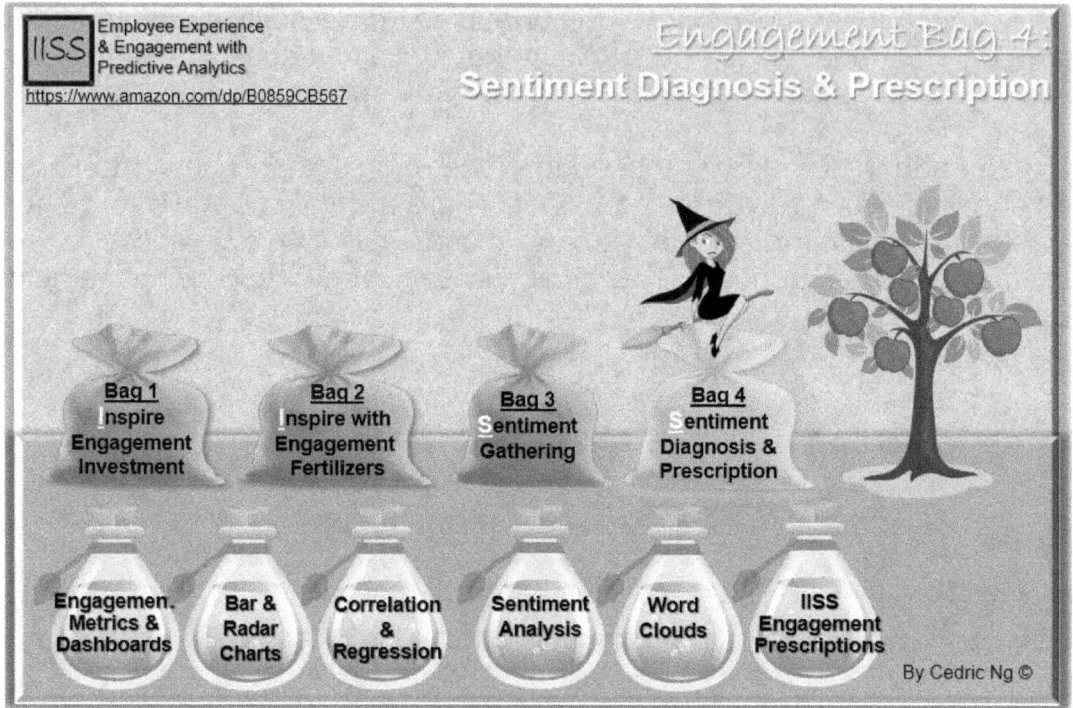

In IISS, **"Sentiment Diagnosis & Prescription"** is the fourth engagement bag because we can only analyse employee sentiment after we have gathered them. There are various ways to analyse employee sentiment. This section covers these approaches to analyse employee sentiment:
- Engagement Metrics & Dashboards
- Bar & Radar Charts
- Correlation & Regression
- Sentiment Analysis
- Word Clouds
- IISS Engagement Prescriptions

Apple tree: You can analyze your apples by visually inspecting it, cutting it, or eating it, and then sharing your observations, analysis and conclusion with others. What's invisible to us, is how the apple tree is, underground and whether the roots are healthy or rotting - we won't know that until problem surfaces (e.g. the tree may shed its leaves, lose its color and vibrancy, weaken its grip on the soil and be uprooted by strong winds). There are usually one or more issues causing the roots to rot. Before the tree can be treated and nursed back to health, we need to determine the root causes.

Organisation: Disengagement is like a sick tree. We are masters of disguise at work and at home. On the surface, disengaged people may be smiling on the outside, but underneath, discontent may be spreading, and they may be suffering inside. However, people can't hide behind the smile forever – disengagement will show. Disengagement occurs over a period of time and slowly eats away at the passion you have for your job until there is none left. There are usually one or more factors causing the disengagement disease. To heal disengagement, you need to examine the roots for the underlying cause. You can analyze employee engagement with Engagement Metrics & Dashboards, Bar Charts, Radar Charts, Word Clouds, Sentiment Analysis, Correlation & Regression. After analysis, you can use the IISS model's list of Engagement Prescriptions to build great employee experience & engagement.

Apple: You can predict how big your apples will be, by the amount of fertilizers that your put. Similarly, you can predict your business results by the change in your employee engagement levels using correlation and regression.

Organisation: Despite your best efforts, some apples will be sweet and some sour. Similarly, despite your best efforts, some employees will be engaged and some disengaged.

Reference
(1) romyantoine (2018) Employee Engagement https://onest g/blog/job-satisfaction-vs-employee-engagement-equal/ 1 October 2019
(2) Robyn Reilly (2019) Five Ways to Improve Employee Engagement Now https://www.gallup.com/workplace/231581/five-ways-improve-employee-engagement.aspx (2 October 2019)
(3) Charles Rogel (2018) Employee Satisfaction vs. Employee Engagement in 2018 https://decision-wise.com/job-satisfaction-vs-employee-engagement/ 1 October 2019

1.2 What is Employee Experience & Engagement?

Employee engagement does not mean employee satisfaction. While company game rooms and parties are fun - making employees satisfied or happy, is different from making them engaged. Someone might be happy at work, but that doesn't necessarily mean they are working hard and productively for the organization.

Satisfaction versus Engagement	
Satisfaction	Engagement
Involve only feelings	Involves feelings but requires action

Myths about Engagement	
Myth	Fact
Engagement is about people working harder	Engagement is about empowering and inspiring employees to give more of their own accord
Engagement is about happy employees	You can't have engagement without happiness, but happiness doesn't equal engagement
The same things engage everyone	People engage differently based on the things that matter to them
Engagement makes leaders redundant	Engagement needs less management but more leadership
You can engage employees with a compelling end goal	Employees become engaged when their goals are important to them in the context of the organization's larger goals.

Employee Satisfaction

Employee Satisfaction is a measurement of an employee's happiness with current job and conditions; but doesn't measure how much effort they're willing to put into their job. A survey question for satisfied employee would ask: Are you going to stay? [1]

Organizations around the world spend a lot of money trying to "drive engagement" - the problem is they are wasting most of that money. The biggest reasons for that failure is their programs are usually addressing employee satisfaction. Engagement and satisfaction are different things. When you install an arcade game in the office, you may temporarily enhance employee satisfaction. But it doesn't last. You can't win hearts with arcade games.

Employee Engagement

Employee Engagement takes into account employee's motivation and the amount of effort they're willing to put in to help the company achieve its goals. It is the secret sauce behind inspired, productive and profitable organizations. Real engagement does not lead to burnout. You can be exhausted from working on a project, but if you are engaged you won't lose your enthusiasm for your work. Employee satisfaction is the minimum requirement for an employee to be genuinely engaged. A survey question for engaged employee would ask: Are you in line with our goals? [1]

Employee Experience

Employee experience is the sum of perceptions of employees about their interactions with the organization.

Employee Life Cycle

Employee Life Cycle is part of Employee Experience. Employee Life Cycle includes all the steps from an employee's first contact with the organization to their departure. It covers HR processes such as recruitment, onboarding, development, promotion, and exit interviews.

Two employees can have similar Employee Life Cycle, but different Employee Experience, because of differences in their expectations. A positive Employee Experience isn't just about what the company throws at the employee (e.g. higher pay, perks) – it is about whether they meet their expectations.

Three types of employees

According to Gallup, there are three types of employees (Engaged, Not-engaged, Actively disengaged). Gallup's global research finds only 13% of employees worldwide are engaged at work. [2]

- **Engaged** employees go the extra mile, work with passion, feel a connection to their company, and move your business forward.
- **Not-engaged** employees are do just enough to fulfil their job requirements – they are a good untapped opportunity for businesses to improve their performance and profitability.
- **Actively disengaged** employees aren't just unhappy at work. They are busy acting out their unhappiness. They undermine what their engage coworkers accomplish.

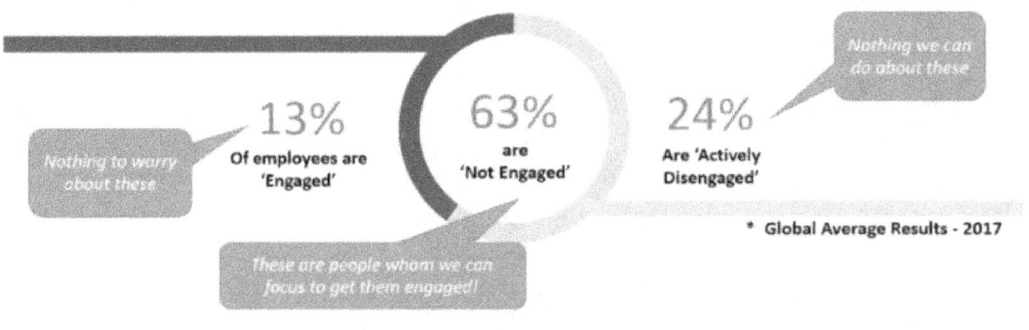

Source: Sareen Babu Madupu (2019) Top 9 Actionable Employee Engagement Ideas and Activities https://acuvate.com/blog/top-9-actionable-employee-engagement-ideas-and-activities/ (3 October 2019)

References
(1) romyantoine (2018) Employee Engagement https://onest g/blog/job-satisfaction-vs-employee-engagement-equal/ 1 October 2019
(2) Robyn Reilly (2019) Five Ways to Improve Employee Engagement Now https://www.gallup.com/workplace/231581/five-ways-improve-employee-engagement.aspx (2 October 2019)

Employee Net Promoter Score (eNPS)

Employee Net Promoter Score (eNPS) is an approach for companies to measure employee loyalty. Employee Net Promoter Score measures the probability your employee will recommend your company as a place to work and the products/services sold. Example of eNPS question: *Using a scale of 0 to 10, how likely will you recommend this company as a place to work?*

Fred Reichheld, a researcher from Bain & Company, wrote an article in Harvard Business Review about how one question can be used to predict companies' growth, "It turned out that a single survey question can, in fact, serve as a useful predictor of growth.... In fact, in most of the industries that I studied, the percentage of customers who were enthusiastic enough to refer a friend or colleague...correlated directly with differences in growth rates among competitors". [1]

Reichheld group the answers were in three groups [2]:
- Promoters: score 9 to 10 for the question and are very likely to recommend. To get actionable insights, ask the promoters what makes them willing to recommend.
- Passively Satisfied: score 7-8 for the question. They are neutral because they are neither going to promote or talk negatively about the company.
- Detractors: score 0-6 for the question and will not recommend. To get actionable insights, ask the detractors the reason for the score they gave.

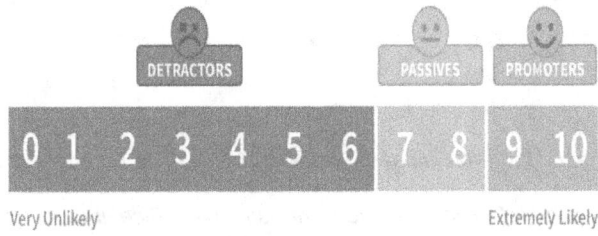

Source: Jacob Shriar, Employee Net Promoter Score, https://www.officevibe.com/employee-engagement-solution/employee-net-promoter-score (2 November 2018)

Formula to Calculate The eNPS Score:

$$eNPS = \% \text{ Promoters} - \% \text{ Detractors}$$

If the results you get from an eNPS survey is: 30 Detractors, 50 Passives, 10 Promoters,

$$eNPS = 10\%(\text{Promoters}) - 30\%(\text{Detractors}) = -20\%$$

People who scored 7 or 8 are not included in the computation because they are neutral. As the eNPS score is represented as a number, remove the percentage sign and your eNPS is -20. Keep in mind that the real value is not so much not the score, but the written feedback from employees about what you company can improve in. Here some initiatives that you can use to improve your eNPS score, and ultimately your business results:

- Communicate your results and plan – Employees will be frustrated if you conduct a survey and don't share the results of the survey and don't follow up with a plan for implementing changes.

- Analyzing the data – To be able to see trends, you can group your data in various ways: age groups, gender, length of service, department.

Reference
(1) Frederick F. Reichheld, Harvard Business Review (2003), The One Number You Need to Grow (2 November 2018)
(2) Jacob Shriar, Employee Net Promoter Score, https://www.officevibe.com/employee-engagement-solution/employee-net-promoter-score (2 November 2018)

1.3 Climate vs Culture vs Engagement

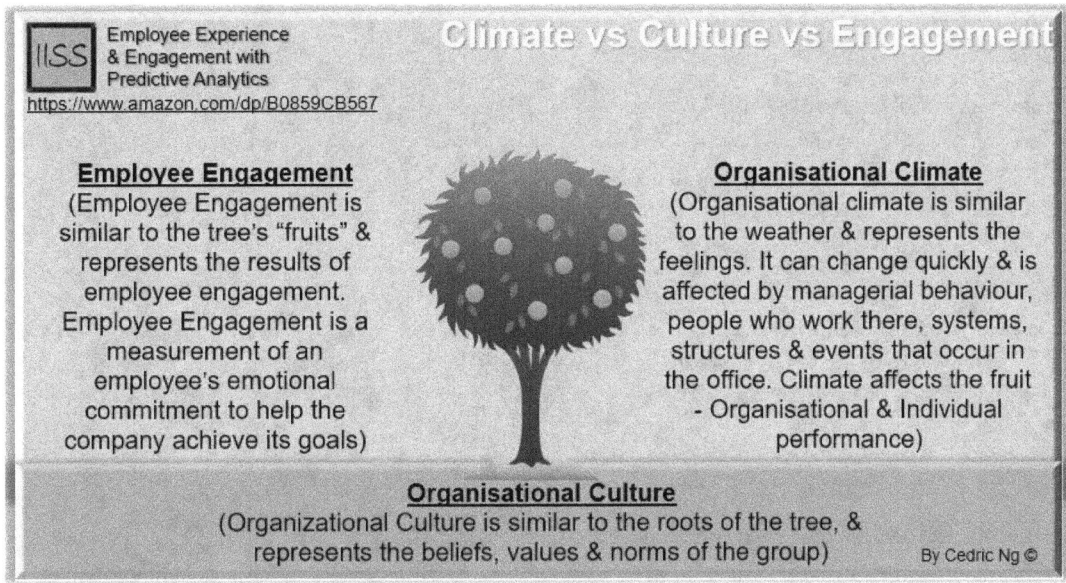

It's easy to get culture and climate confused, as they are intertwined, and one tends to impact and feed off the other. Climate can change quickly due to an event, while culture takes time to develop. Organisational Climate, Culture, and Employee Engagement is like a tree:

- **Organizational Climate:** Organisational climate is similar to the weather and represents the feelings. It can change quickly, and is affected by the managerial behaviour, the people who work there, systems, structures and events that occur in the office. Climate affects the fruit - Organisational & Individual performance.

- **Organizational Culture**: Organizational Culture is similar to the roots of the tree, and represent the enduring beliefs, values and norms of the group.

- **Employee Engagement:** Employee Engagement is similar to the tree's "fruits" and represents the results of employee engagement. Employee Engagement is a measurement of an employee's emotional commitment to help the company achieve its goals.

2.0 Engagement Bag 1: Inspire Engagement Investment

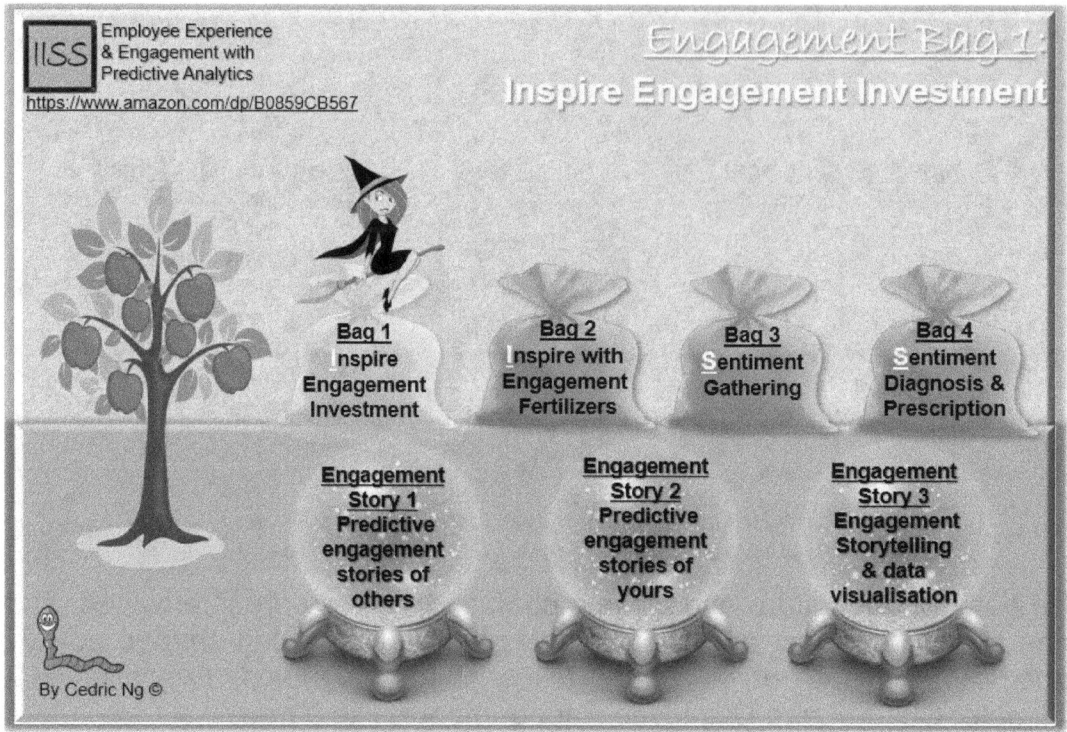

In IISS, **"Inspire Engagement Investment"** is the first engagement bag because the organization and managers need to know how engagement benefits them, before they support it. This section covers how to Inspire Engagement Investment with "3 Types of Engagement Stories":
Engagement Story 1) Predictive engagement stories of others
Engagement Story 2) Predictive engagement stories of yours
Engagement Story 3) Engagement storytelling & data visualisation

Apple farm: To convince organizations to invest in your apple farm, you need to show them how sweet the apples can be, if they invest in fertilizers.

Organization: Similarly, to convince organizations to invest in employee engagement programs, you need to show them how organizational objectives can be met, if they invest in employee engagement programs. Despite what businesses say in their mission statements, they don't exist just to "care for our employees' well-being". Ultimately, most businesses care about engagement only if it helps them make money. That's why a significant portion of this book is devoted to show how engagement improves people's performance and business results.

Apple: To convince people to buy your apples, you need to show them how sweet the apples can be, if they take care of their apple tree.

Managers: Similarly, to convince managers to engage their staff, you need to show them how their team objectives can be met, if they take care of their employees.

2.1 Engagement Story 1: Predictive engagement stories of others

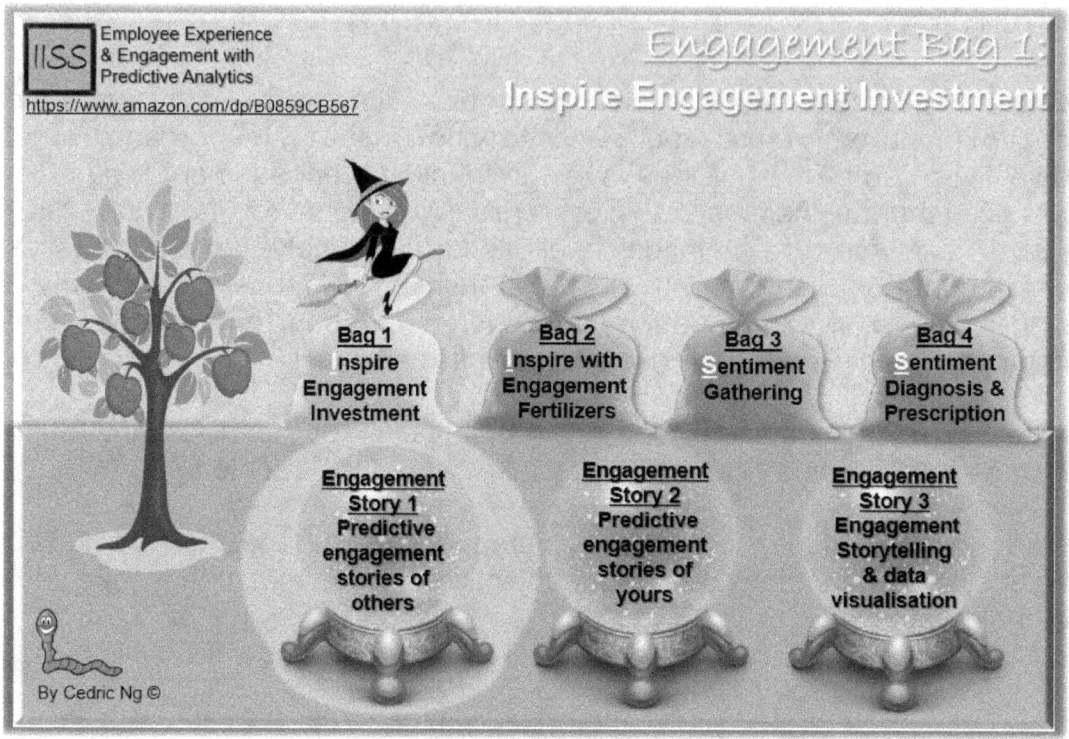

Numerous researches have shown that there is a relationship between Employee Engagement and variables such as: employee demographics, employee churn, absenteeism, employees' health, inventory shrinkage, sales, safety, profit, innovation, customer satisfaction, and total shareholder return.

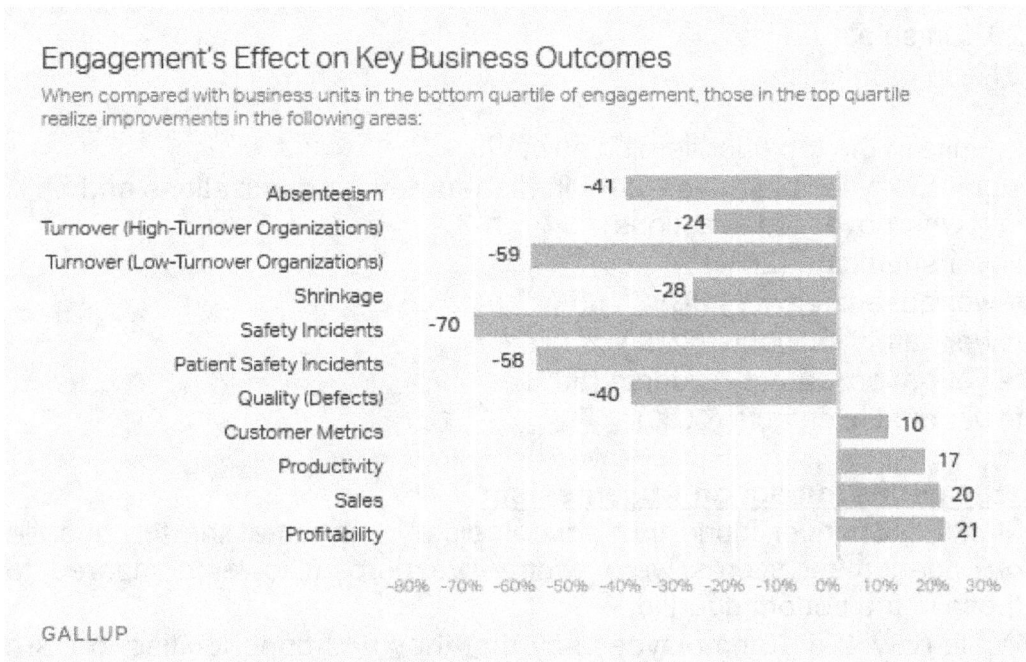

Source: Robyn Reilly (2019) Five Ways to Improve Employee Engagement Now https://www.gallup.com/workplace/231581/five-ways-improve-employee-engagement.aspx (2 October 2019)

Gallup researchers studied the differences in performance between engaged and actively disengaged work units and found that work units in the top quartile in employee engagement outperformed bottom-quartile units by: [1]
- 10% on customer ratings
- 17% in productivity
- 20% in sales
- 21% in profitability

Work units in the top quartile also saw: [1]
- significantly less turnover (24% in high-turnover organizations and 59% in low-turnover organizations)
- fewer shrinkage (28%)
- fewer absenteeism (41%)
- fewer safety incidents (70%)
- fewer patient safety incidents (58%)
- fewer quality defects (40%)

Engagement's impact on Absenteeism
- Marks & Spencer found that absenteeism in stores at the top quartile of engagement scores were twenty-five percent lower compared to those in the bottom quartile. [2]
- While only 4 in 10 employees say that they had opportunities to learn and grow, Gallup suggests that improving that ratio to 8 in 10 can allow your business to achieve a 44% drop in absenteeism and a 16% jump in productivity. [3]

Engagement's impact on Employee Attrition
- Corporate Leadership Council found that engaged employees are eighty-seven per cent less likely to resign. [4]
- Standard Chartered found their branches with high employee engagement had 46% lower voluntary turnover. [5]

Employee Demographics' impact on Engagement
- Shoshana Dobrow Riza (London School of Economics and Political Science), Yoav Ganzach (Tel Aviv University), and Yihao Liu (University of Florida) found that Job satisfaction tends to improve as we get older but also tends to decrease the longer we stay at a particular job. [6]
- Eric van Duin discovered that a manager's tenure strongly correlated to its team's engagement at PostNL N.V. The shorter the manager's tenure, the higher the engagement of the team. With the findings, HR reviewed their training programme for managers with long tenures, and as a result improved their engagement with their teams. [7]

Engagement's impact on Health
- Engaged employees take an average of 2.7 sick days per year, whereas disengaged employees take 6.2 sick days per year. [8]

Engagement's impact on Innovation
- Krueger & Killham found that fifty-nine percent of engaged employees indicated that their job 'brings out their most creative ideas', compared to only three percent for disengaged employees. [9]

Engagement's impact on Inventory shrinkage
- Research showed that low employee engagement levels are linked to increased inventory shrinkage. Inventory shrinkage refers to the loss of inventory. For example, if the inventory records show that there are 5,520 units of Product ABC, but a physical count show that there are only 5,510 units, there is an inventory shrinkage of 10 units. Inventory shrinkage can be due to employee theft, shoplifting, damage, etc. The MacLeod Review cited a meta-analysis that found inventory shrinkage was fifty-one percent worse in workplaces with low engagement compared to those with engaged employees. As an example, if the number of engaged employees increase by five percent in a business with $10M inventory shrinkage, the company might save $250,000. [10]

Engagement's impact on Profitability

- Standard Chartered found that branches where employee engagement was high has sixteen percent higher profit margin growth compared to branches where employee engagement was low. [11]
- A study by ISS found that profitability is high, when employee engagement and customer advocacy are high. The average profitability in countries scoring highest on both employee engagement and customer advocacy was 7.75 percent, versus 4.52 percent for the lowest scoring groups. ISS tested this link by combining selected questions from their Employee Engagement Survey and their Customer Experience Survey with their profitability. In the diagram below, the eNPS (employee net promoter score) and cNPS (customer net promoter score) scores are depicted on the X and Y axis respectively, while the average profitability is shown in the boxes. The figure shows that if the employees and the customers report a high level of satisfaction (as measured by the NPS), then the country will have significantly higher margins than if either of the two scores are low. Implication of this study is that companies should reinvest more in customer service improvements, employee training, on-boarding processes, and service innovation. [12]

Link between margins and employee & customer net promoter scores at a country level

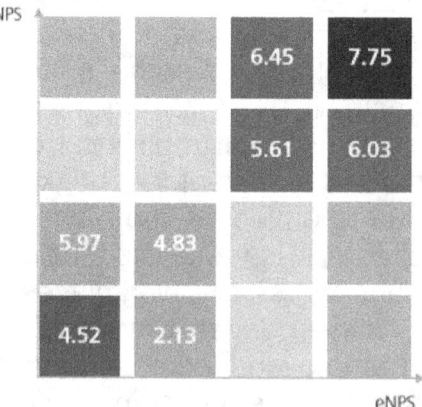

Source: ISS Prospectus, 2014

Source: Morten Kamp Andersen (proacteur), Simon Svegaard (ISS) & Peter Ankerstjerne (ISS). (2015), Linking Customer Experience with Service Employee Engagement, ISS White Paper, pp 10-11.

Engagement's impact on Quality
- DDI studied the employee engagement in two hundred organizations, and found that employees with higher engagement scores have lower Quality errors. In a Fortune 100 manufacturing company, Quality errors as measured by external and internal parts per million was 5,658 for the low-engagement group and only 52 for the high-engagement group! [13]

Engagement's impact on Safety
- MolsonCoors found that engaged employees were five times less likely than non-engaged employees to have a safety incident. [14]

Engagement's impact on Sales
Research has shown that Employee Engagement affects Sales.
- A company featured in the 2016 Engage for Success report reported that engaged salespeople sold an average of $70,000 more per year, compared to disengaged salespeople - which is around 7% extra sales revenue per engaged employee. [15]
- Standard Chartered Bank found that those branches with highly engaged employees produced twenty percent higher returns compared to branches with lower engagement scores. [16]
- Marks and Spencer found that one percent improvement in their employee engagement produced three percent increase in sales per square foot. [16]
- JCPenney found that stores with top engagement scores produced around ten percent more sales per square foot and have thirty six percent higher operating income than stores with low engagement scores. [16]
- Best Buy (a leader in HR predictive analytics) can accurately predict how employee engagement impacts the performance of their stores. A 0.1% increase in employee engagement results in an increase of over $100,000 in the store's annual income. The enormous impact of engagement prompted Best Buy to make its engagement surveys quarterly instead of annually. [17]

Engagement's impact on Service

- A study by ISS found that Employee Engagement correlates strongly with Customer Experience. The strength of this correlation was 0.55. Diagram below shows the customer net promoter score (cNPS) and the employee net promoter score (eNPS). Net promoter score (NPS) is a tool to measure the loyalty of a firm's customers and employees. The NPS was based on the question: "How likely is it that you would recommend a company, product or service to a friend or colleague?" This answer was scored based on a 0-10 scale. Promoters (loyal enthusiasts) are those who respond with a score of 9 or 10. Passives are those who respond with a score of 7 or 8. Detractors (unhappy customers) are those who respond with a score of 0 to 6. Calculate the NPS by subtracting the percentage who are detractors from the percentage who are promoters. The drivers behind customer experience are: Motivation and engagement of service staff, Amount of training and quality of service staff, and the ability to act on customer expectations by customer service staff. With the findings, ISS recommendations were to monitor metrics on service employee engagement in areas such as the eNPS (employee net promoter score), employee churn, absenteeism and training hours. [18]

Correlation between eNPS and cNPS

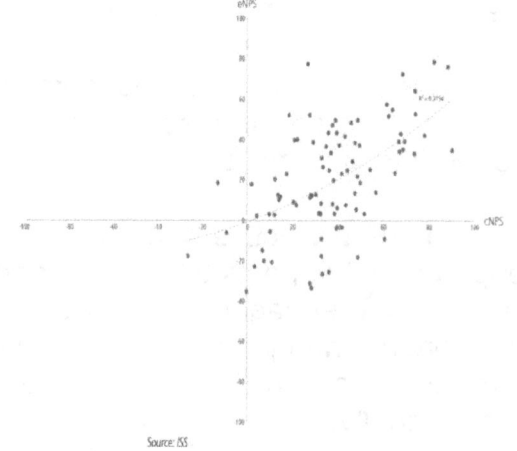

Source: Morten Kamp Andersen (proacteur), Simon Svegaard (ISS) & Peter Ankerstjerne (ISS). (2015), Linking Customer Experience with Service Employee Engagement, ISS White Paper.

Engagement's impact on Total Shareholder Returns (TSR)
- A study conducted across 39 organizations showed that organizations with highly engaged employees achieve seven times greater 5-year total shareholder return (TSR) than organizations whose employees are less engaged. In companies where 60 to 70 percent of employees were engaged, average total shareholder's return (TSR) is 24.2 percent. in companies with only 49 to 60 percent of their employees engaged, TSR fell to 9.1 percent; companies with engagement below 25 percent suffered negative TSR. [19]

Leadership's impact on Engagement
- Gallup's research in the US shows that the actions of leaders account for up to 70% variance in employee engagement scores. [20]

References

(1) Robyn Reilly (2019) Five Ways to Improve Employee Engagement Now https://www.gallup.com/workplace/231581/five-ways-improve-employee-engagement.aspx (2 October 2019)

(2) Culture Amp (2018) Why Employee Engagement Matters. https://www.cultureamp.com/resources/guides/foundation-guides/why-employee-engagement-matters.html#ref-3 (14 November 2018)

(3) Tanvir Haque (2018) Four Strategies To Boost Employee Engagement That Can Improve Your Bottom Line. https://www.entrepreneur.com/article/312697 (20 November 2018)

(4) Corporate Leadership Council, Corporate Executive Board. Driving Performance and Retention through Employee Engagement: a quantitative analysis of effective engagement strategies. 2004. Cited in MacLeod 2009, p.14.

(5) MacLeod, D., and Clarke, N. Engaging for Success: Enhancing Performance Through Employee Engagement—A Report to Government. London: Department for Business Innovation and Skills. 2009. Crown copyright.

(6) Association for Psychological Service (2018) Job Satisfaction Tends to Increase with Age https://www.psychologicalscience.org/news/minds-business/job-satisfaction-tends-to-increase-with-age.html (21 November 2018)

(7) Nigel Guenole, Jonathan Ferrar, and Sheri Feinzig (2017). The Power of People, Pearson FT Press.

(8) Harter, J.K., Schmidt, F. L., Kilham, E. A., Asplund, J.W., (2006), Gallup Q12 Meta-Analysis. Cited in MacLeod, 2009, p. 36.

(9) Krueger, J. & Killham, E 'The Innovation Equation.' Gallup Management Journal, 2007. Cited in MacLeod 2009, p. 12.

(10) Culture Amp (2018) Why Employee Engagement Matters. https://www.cultureamp.com/resources/guides/foundation-guides/why-employee-engagement-matters.html#ref-3 (14 November 2018)

(11) Culture Amp (2018) Why Employee Engagement Matters. https://www.cultureamp.com/resources/guides/foundation-guides/why-employee-engagement-matters.html#ref-3 (14 November 2018)

(12) Morten Kamp Andersen (proacteur), Simon Svegaard (ISS) & Peter Ankerstjerne (ISS). (2015), Linking Customer Experience with Service Employee Engagement, ISS White Paper, pp 10-11.

(13) Richard S. Wellins, Paul Bernthal (2015) Employee Engagement: The Key to Realizing Competitive Advantage. DDI. https://www.ddiworld.com/ddi/media/monographs/employeeengagement_mg_ddi.pdf?ext=.pdf (20 November 2018)

(14) Kevin Kruse (2012) Why Employee Engagement? (These 28 Research Studies Prove the Benefits) https://www.forbes.com/sites/kevinkruse/2012/09/04/why-employee-engagement/#36bbedf63aab (20 November 2018)

(15) Court-Smith, J., 'The Evidence: Case Study Heroes and Engagement Data Daemons', Engage for Success, April 2016, p. 23.

(16) Parent, J. D., & Lovelace, K. J. (2015). The Impact of Employee Engagement and a Positive Organizational Culture on an Individual's Ability to Adapt to Organization Change.2015 Eastern Academy of Management Proceedings: Organization Behavior and Theory Track, 1-20. (14 November 2018)

(17) Mohit Sharma, Talent Analytics: From Buzzword to Reality (2018), https://sightsinplus.com/2018/05/25/talent-analytics-from-buzzword-to-reality/ (September 2018)

(18) Morten Kamp Andersen (proacteur), Simon Svegaard (ISS) & Peter Ankerstjerne (ISS). (2015), Linking Customer Experience with Service Employee Engagement, ISS White Paper, pp 3-9.

(19) Kevin Kruse (2012) Why Employee Engagement? (These 28 Research Studies Prove the Benefits) https://www.forbes.com/sites/kevinkruse/2012/09/04/why-employee-engagement/#36bbedf63aab (20 November

(20) Tanvir Haque (2018) Four Strategies To Boost Employee Engagement That Can Improve Your Bottom Line. https://www.entrepreneur.com/article/312697 (20 November 2018)

2.2 Engagement Story 2: Predictive engagement stories of yours

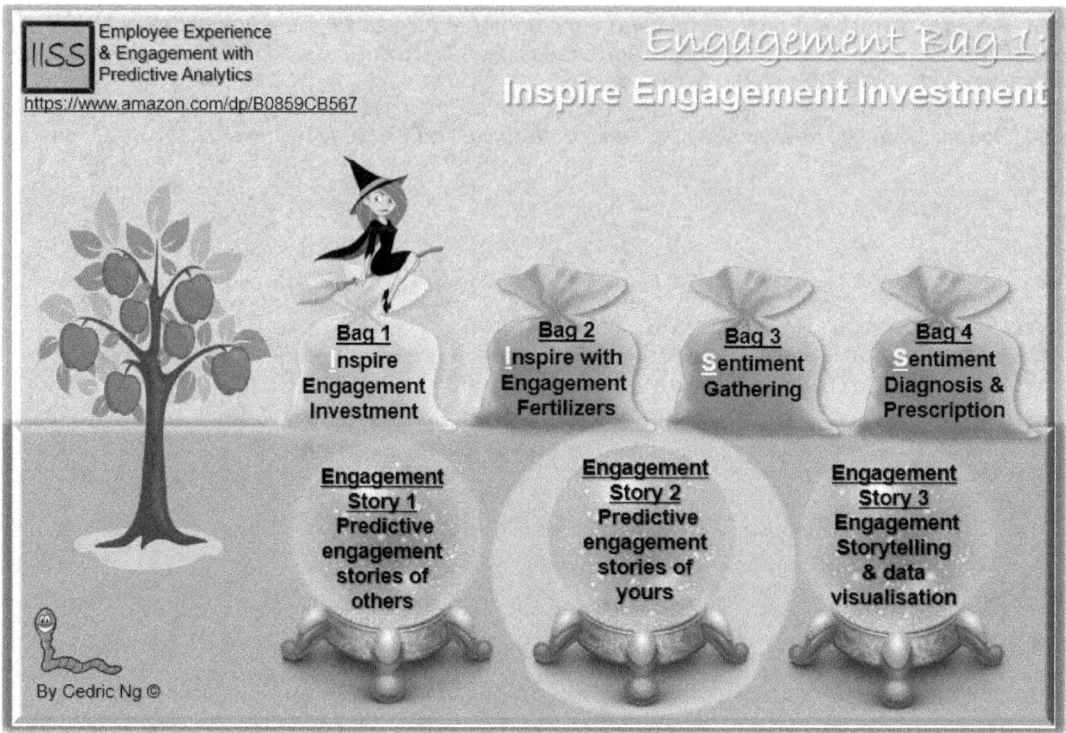

2.2.1 Analytics Maturity Model

Human Resources Professionals have been talking about HR analytics for years, but 2016 saw the biggest leaps toward true HR analytics capabilities. Organizations are progressing from the most fundamental HR data analysis stages of "descriptive and diagnostic analytics" to the more complicated "predictive and prescriptive analytics". According to the "Global Human Capital Trends 2016" report by Deloitte, the percentage of companies that believe they are capable of developing predictive models doubled from four percent in year 2015 to eight percent in year 2016.

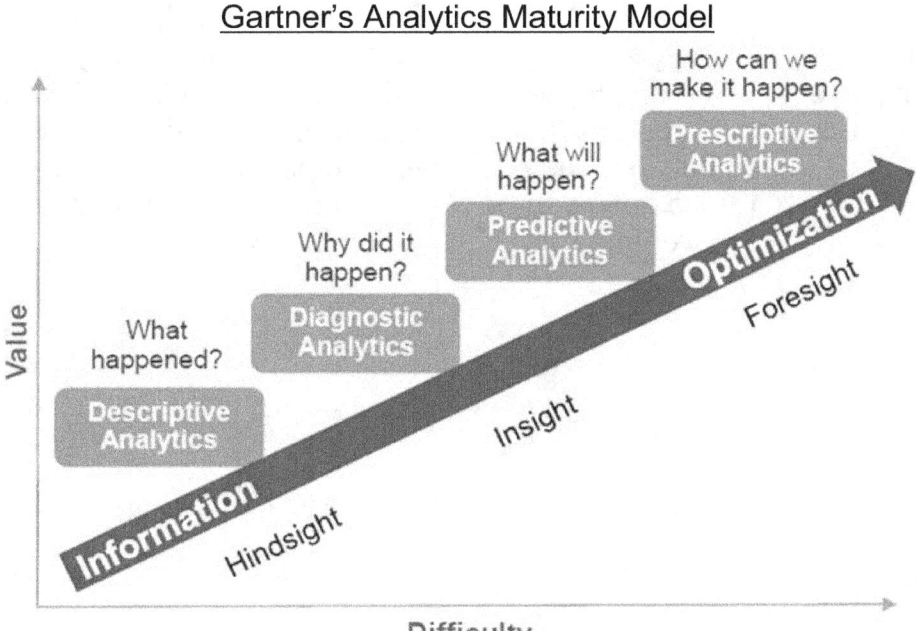

Gartner's Analytics Maturity Model

Source: https://medium.com/taras-kaduk/4-stages-of-data-analytics-maturity-challenging-gartners-model-590eb5ebe6d1

In the Analytics Maturity Model, there are four levels of data analytics used across all industries, and the increasing value each step provides to your organization:
- Level 1: Descriptive Analytics
- Level 2: Diagnostic Analytics
- Level 3: Predictive Analytics
- Level 4: Prescriptive Analytics

Fours Levels of Gartner's Analytics Maturity Model

Level 1 Descriptive Analytics	Level 2 Diagnostic Analytics	Level 3 Predictive Analytics	Level 4 Prescriptive Analytics
100 employees resigned	100 employees resigned because they are not given overtime pay	Giving employees overtime pay will improve retention by x%	At least 2 hours of overtime will improve retention

Descriptive and diagnostic analytics are commonly used, while predictive and prescriptive analytics is less common. Progressing from descriptive analytics towards diagnostic, predictive and prescriptive analytics requires much more technical ability (higher difficulty), but also uncovers more insights (value).

Level 1: Descriptive Analytics

The first level of data analytics is descriptive analytics. Descriptive analytics is at the foundation and most commonly used of data analytics. Descriptive analytics answers "what happened" by describing or summarizing raw data usually in the form of dashboards. Most the basic statistics (sums, averages, count) that we use fall into this category. Descriptive statistics are useful to show things like, Key Performance Indicators (KPI's), dashboards, profit per region, how many employees joined/left, total training hours. HRIS is typically used to obtain descriptive insights about employees as it reports what happened and what is currently happening. HRIS provides information about your employee demographics, characteristics, costs and performance. However, HRIS systems do not have the capability of discovering patterns and the correlations in these patterns.

Level 2: Diagnostic Analytics

The second level of data analytics is diagnostic analytics. After asking "what happened" you may dive deeper and ask "why did it happen"? Diagnostic analysis uses the data that was summarized in descriptive analytics and analyses further to find the cause of that result. It answers the questions raised in Descriptive Analytics. Examples include: why did sales go down, why are shipments slow.

Level 3: Predictive Analytics

The third level of data analytics is Predictive analytics. Predictive analytics tries to answer the question "what is likely to happen". It uses HRIS to gain descriptive insights about employees, and then use Predictive HR Analytics to build predictive models. Predictive analysis uses the insight from Diagnostic analytics to predict of the results of various variables with statistical modeling. Uses of predictive analytics include: predicting employee churn, predicting sales, etc. Predictive analytics answers questions such as: Which employees are most likely to leave in the next six months? Which employees are most likely to perform?

Level 4: Prescriptive Analytics

The fourth level of data analytics is Prescriptive analytics. Prescriptive analytics prescribe what action to take to remove a future problem or take advantage of a promising trend. It combines the insight from all previous analyses to ascertain the course of action to take. Whilst predictive analytics forecasts for what might happen, prescriptive analytics recommends one or more courses of action along with its predicted future scenarios, allowing companies to assess several possible outcomes based upon their actions. Examples of prescriptive analytics include: which customer segment shall we target next year to improve profitability.

Descriptive Analytics versus Predictive Analytics

The easiest way to understand the differences between descriptive and predictive HR analytics is to look at the answers they generate. Descriptive analytics answer questions about the demographics and characteristics of your employees. Whereas, predictive HR analytics go beyond the descriptions generated by descriptive HR analytics by providing predictive answers and prescribing specific actions or recommendations.

For most descriptive HR analytics, predictive HR analytics questions can be developed as you can see in the list below.

Descriptive Analytics	Predictive Analytics
How long has John been with the company	How long will john stay with the company
What is John's marital status, salary market-ratio, performance rating, traveling time to office?	What variables affect John's tenure with the company?
Who are our best performing call center employee?	Which of our call center employees are likely to be best performers?
How many hours of training did our Sales employees clock last year	What is the impact of the training programme on Sales?
How many Sales employees leave last quarter?	Which Sales employee is likely to leave next quarter?
What is the engagement score?	What impact does engagement scores have on accident rates?

There are two approaches to predict the impact of employee engagement:
- Correlation
- Regression

2.2.2 Correlation analysis of engagement investment

In this example, we use correlation to find out whether Staff Engagement Index is a good predictor of business concerns such as Customer Satisfaction, Revenue per Headcount, and No. of Workplace Accidents.

Correlation can be used to help understand what drives employee engagement, and how employee engagement affects business concerns such as customer satisfaction, revenue, and workplace accidents.

	A	B	C	D	E
	Year	Staff Engagement Index	Customer Satisfaction Index	Revenue per Headcount	No. of Workplace Accidents
2	1	50%	60%	500	2
3	2	67%	70%	800	4
4	3	39%	40%	400	1
5	4	54%	55%	500	2
6	5	80%	70%	750	5
7	6	53%	50%	600	1
8	7	45%	50%	550	1
9	8	70%	68%	700	3
10	9	69%	65%	640	3
11	10	63%	70%	720	4
12	11	60%	50%	600	2

1) Install "Analysis ToolPak", an Excel add-in

"Analysis ToolPak" is an add-in for Microsoft Excel that comes with Microsoft Excel. To be able to run regression using Excel, you need to first install "Analysis ToolPak", an Excel add-in program that provides data analysis tools. To load the Analysis ToolPak add-in, follow these steps:

- On the File tab, click Options.

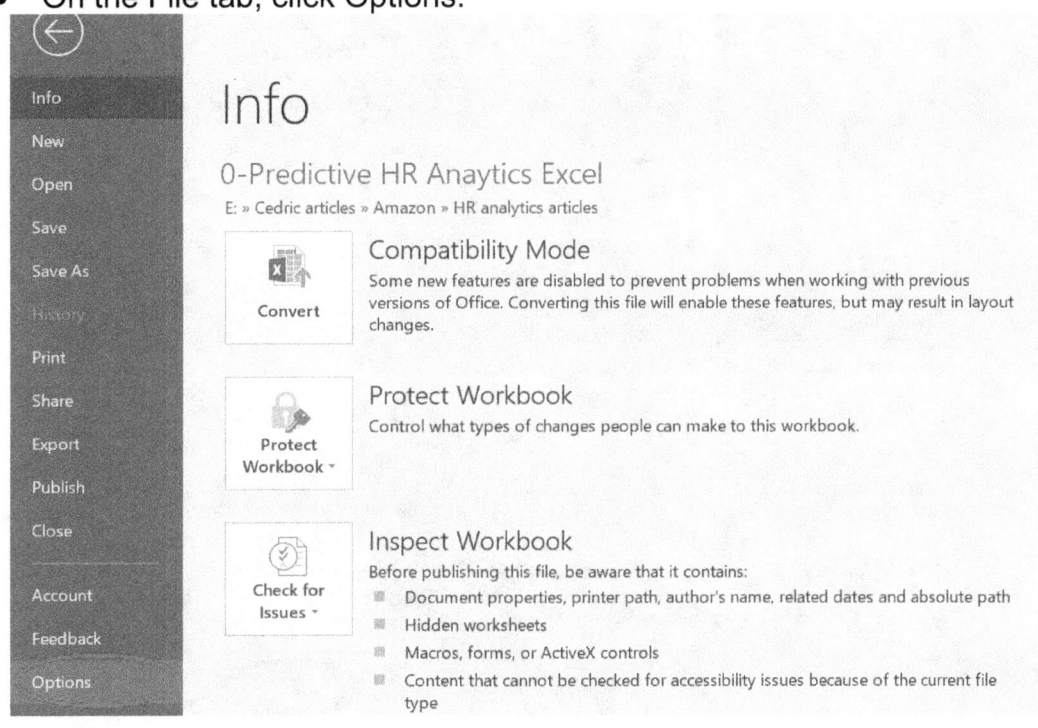

- Under Add-ins, click Analysis ToolPak and click the "Go" button.

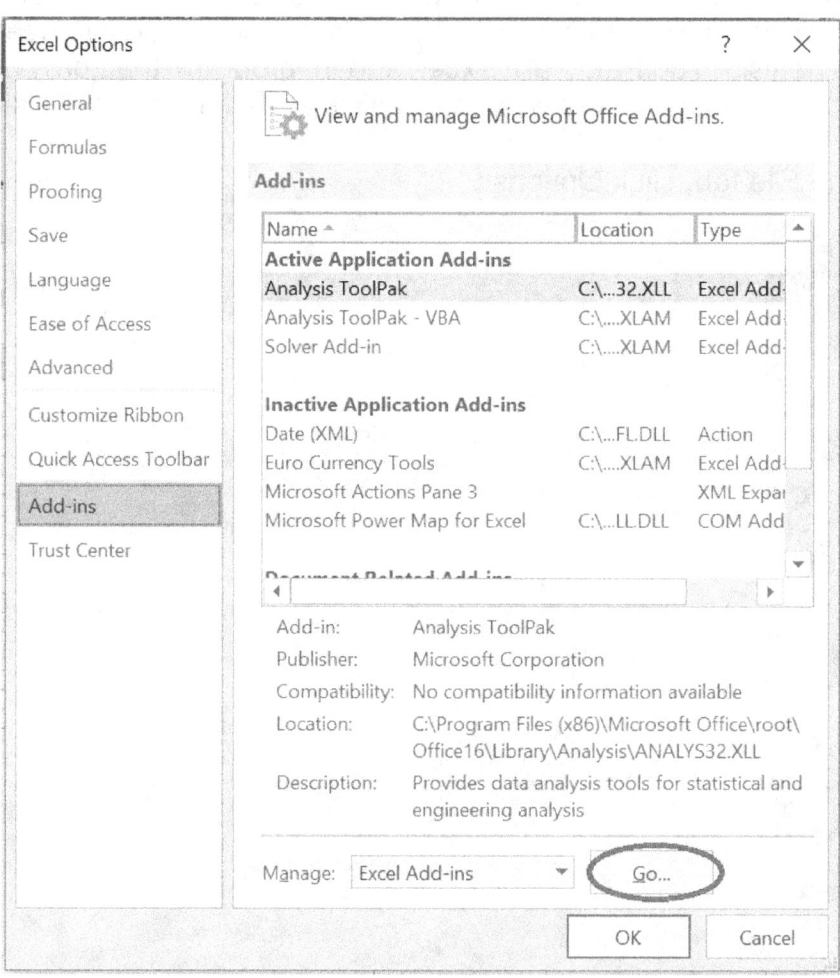

- Click "Analysis ToolPak" and click on OK.

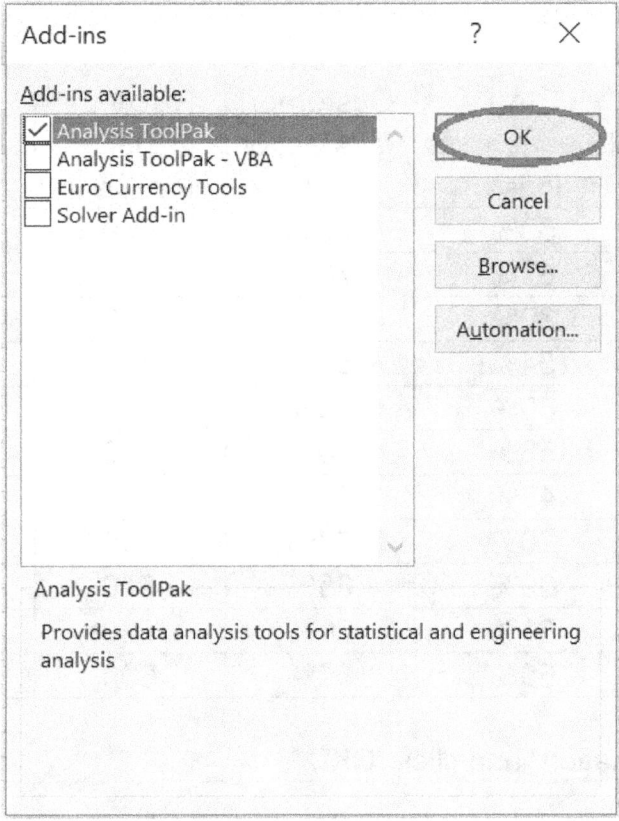

- On the Data tab, in the Analysis group, you are now able to click on "Data Analysis".

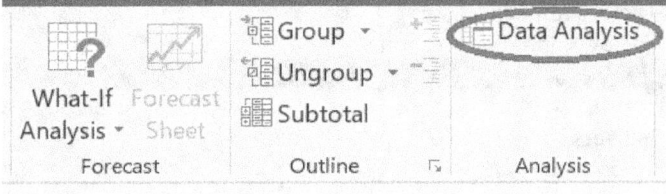

2) Copy the example data in the following table, and paste it in cell A1 of a new Excel worksheet.

	A	B	C	D	E
1	Year	Staff Engagement Index	Customer Satisfaction Index	Revenue per Headcount	No. of Workplace Accidents
2	1	50%	60%	500	2
3	2	67%	70%	800	4
4	3	39%	40%	400	1
5	4	54%	55%	500	2
6	5	80%	70%	750	5
7	6	53%	50%	600	1
8	7	45%	50%	550	1
9	8	70%	68%	700	3
10	9	69%	65%	640	3
11	10	63%	70%	720	4
12	11	60%	50%	600	2

3) Select "Correlation" and click "OK".

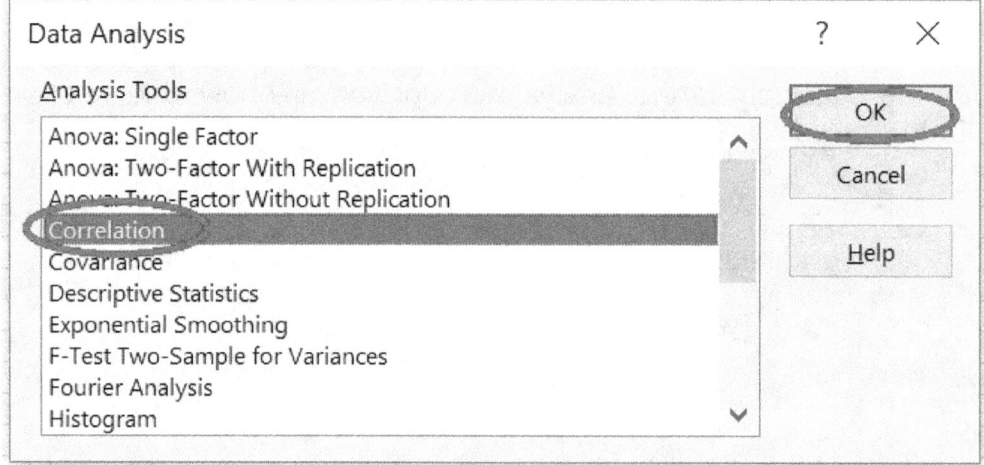

4) After you click OK in the "Data Analysis" dialog box, you will see a "Correlation" dialog box.
5) For "Input Range", select cells (B1:E12).
6) Check "Labels in first row"
7) For "Output Range", select cells (A14).
8) Click "OK"

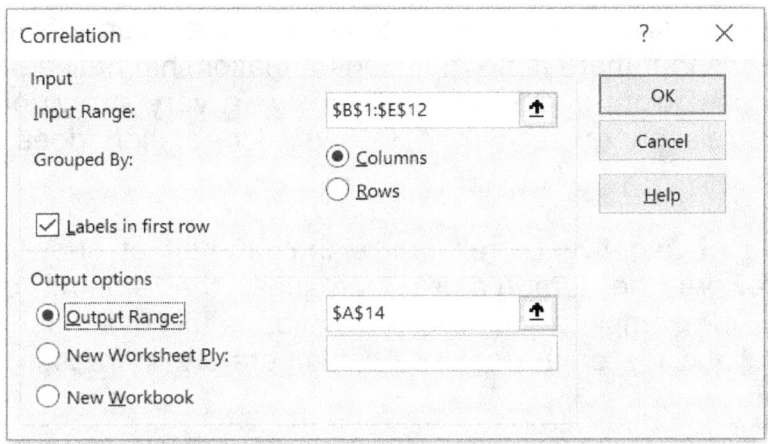

9) After you click "OK", Excel generates the following Correlation analysis.

	Staff Engagement Index	Customer Satisfaction Index	Revenue per Headcount	No. of Workplace Accidents
Staff Engagement Index	1			
Customer Satisfaction Index	0.85	1		
Revenue per Headcount	0.86	0.85	1	
No. of Workplace Accidents	0.89	0.90	0.83	1

A negative correlation coefficient means that an increase in X is associated with a decrease in Y. Similar to a positive correlation, a negative correlation shows a connection between two variables, and the relative strengths are the same. In other words, a correlation coefficient of 0.85 has the same strength as a correlation coefficient of -0.85. Correlation coefficients are always values between -1 and 1, where "-1" means that there is a perfect linear negative correlation, while "1" shows a perfect linear positive correlation. A correlation coefficient of zero, or near to zero, means that there is no meaningful relationship between variables. Correlation coefficient of 0.91 or -0.92 shows a very strong positive and negative correlation respectively. However, correlation does not mean causation.

An example of negative correlation is the amount of snowfall and the temperature. As the temperature increases, the amount of snowfall decreases. An example of positive correlation is the relationship between temperature and ice cream sales. As temperature increases, so do ice cream sales.

10) From the Excel Correlation analysis, Staff Engagement Index is a good predictor of Customer Satisfaction, Revenue per Headcount, and No. of Workplace Accidents, as they have strong correlation of below -0.75 and above 0.75:
- **Customer Satisfaction Index**: correlation of 0.85 with Staff Engagement Index
- **Revenue per Headcount:** correlation of 0.86 with Staff Engagement Index
- **No. of Workplace Accidents:** correlation of 0.89 with Staff Engagement Index

	Staff Engagement Index	Customer Satisfaction Index	Revenue per Headcount	No. of Workplace Accidents
Staff Engagement Index	1			
Customer Satisfaction Index	0.85	1		
Revenue per Headcount	0.86	0.85	1	
No. of Workplace Accidents	0.89	0.90	0.83	1

2.2.3 Regression analysis of engagement investment

Best Buy (a leader in HR predictive analytics) can predict that a 0.1% increase in employee engagement results in an increase of over $100,000 in the store's annual income. [1]

This section, covers how to predict the impact of changes in employee engagement ratings on revenue.

	A	B	C	D	E
1	Year	Average Engagement Rating	Advertising Expense	No. of Salesperson	Revenue
2	1	9.5	96	8	1400
3	2	9.0	89	7	1300
4	3	8.8	85	4	1210
5	4	8.3	83	3	1160
6	5	7.4	85	2	1100
7	6	6.5	80	3	1050
8	7	5.7	85	2	1000
9	8	4.5	75	2	950
10	9	3.3	69	1	890
11	10	3.1	65	2	780

1) Install "Analysis ToolPak", an Excel add-in

"Analysis ToolPak" is an add-in for Microsoft Excel that comes with Microsoft Excel. To be able to run regression using Excel, you need to first install "Analysis ToolPak", an Excel add-in program that provides data analysis tools. To load the Analysis ToolPak add-in, follow these steps:

On the File tab, click Options.

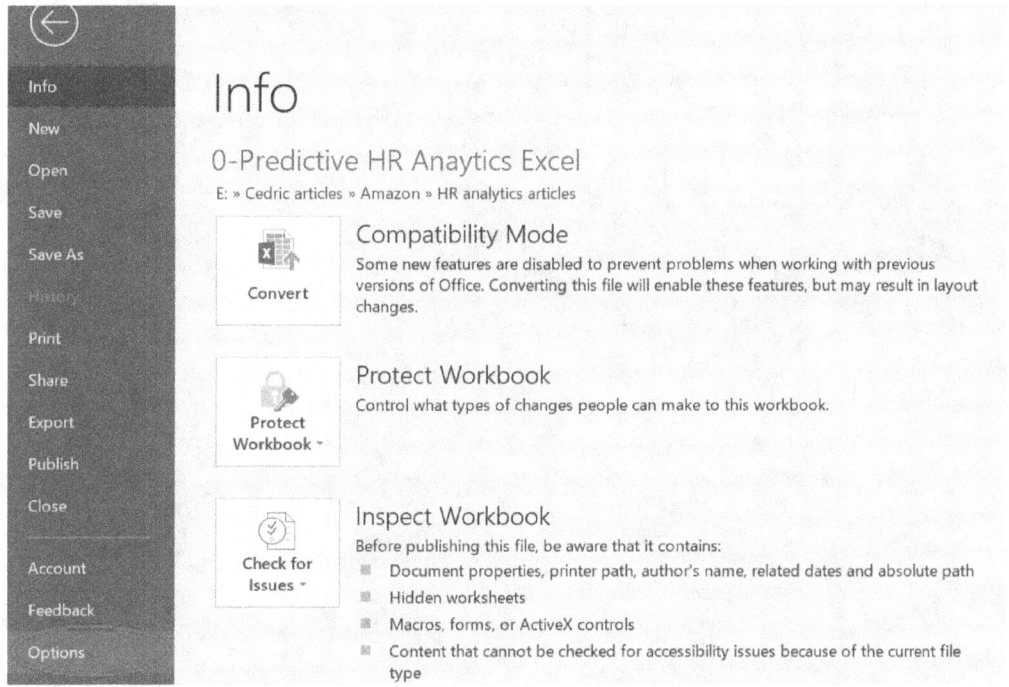

Under Add-ins, click Analysis ToolPak and click the "Go" button.

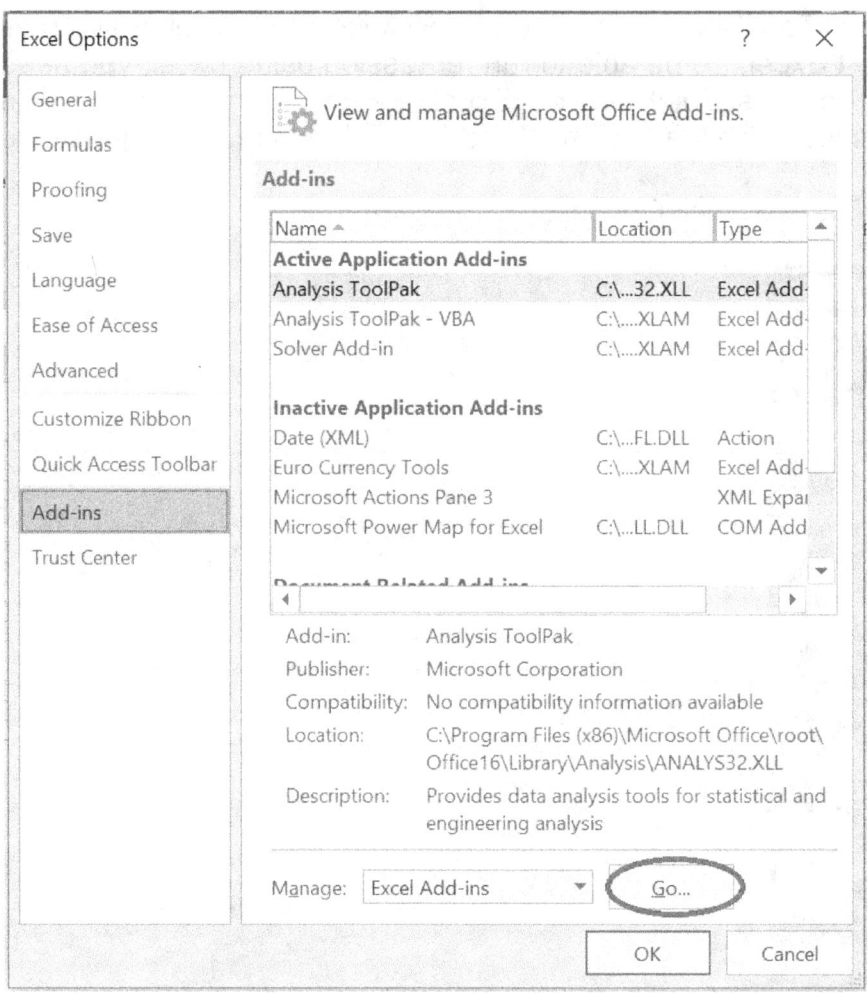

Click "Analysis ToolPak" and click on OK.

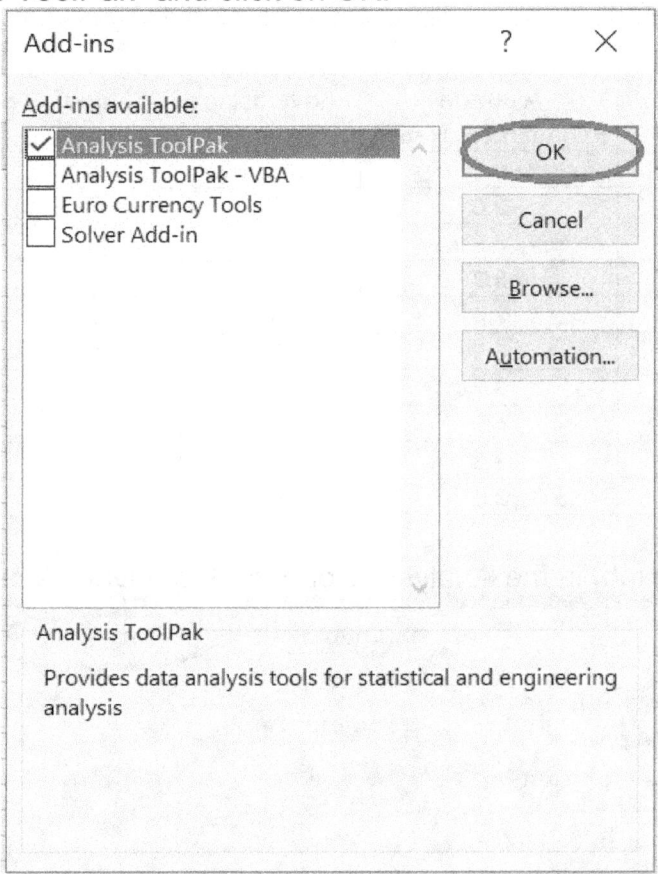

On the Data tab, in the Analysis group, you are now able to click on "Data Analysis".

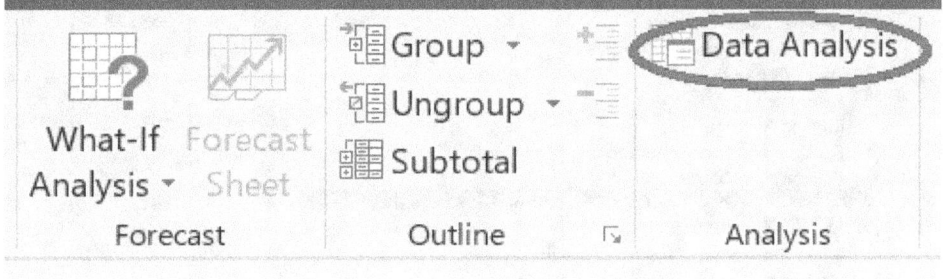

2) Copy the example data in the following table, and paste it in cell A1 of a new Excel worksheet.

	A	B	C	D	E
1	Year	Average Engagement Rating	Advertising Expense	No. of Salesperson	Revenue
2	1	9.5	96	8	1400
3	2	9.0	89	7	1300
4	3	8.8	85	4	1210
5	4	8.3	83	3	1160
6	5	7.4	85	2	1100
7	6	6.5	80	3	1050
8	7	5.7	85	2	1000
9	8	4.5	75	2	950
10	9	3.3	69	1	890
11	10	3.1	65	2	780

3) On the Data tab, in the Analysis group, click on "Data Analysis".

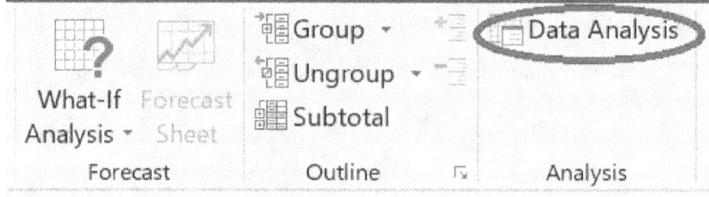

4) Select "Regression" and click "OK".

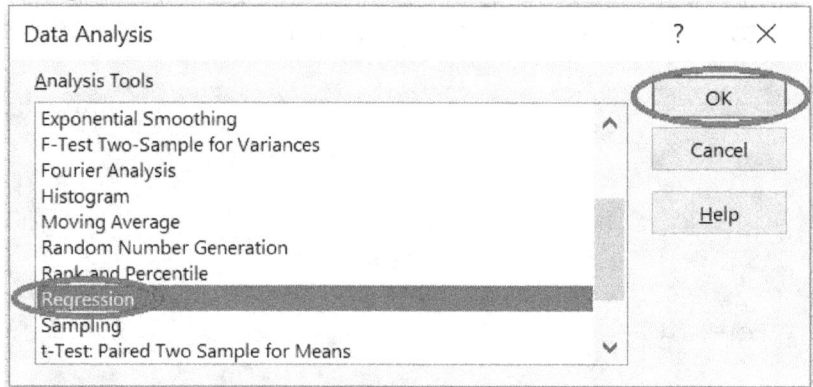

5) After you click OK in the "Data Analysis" dialog box, you will see a "Regression" dialog box.
6) For "Input Y Range", select cells (E1:E11). This is the predictor variable or dependent variable.
7) For "Input X Range", select cells (B1:D11). These are the explanatory variables or independent variables.
8) Check "Labels" box.
9) Click the "Output Range" box, and select cell A13.
10) Click "OK".

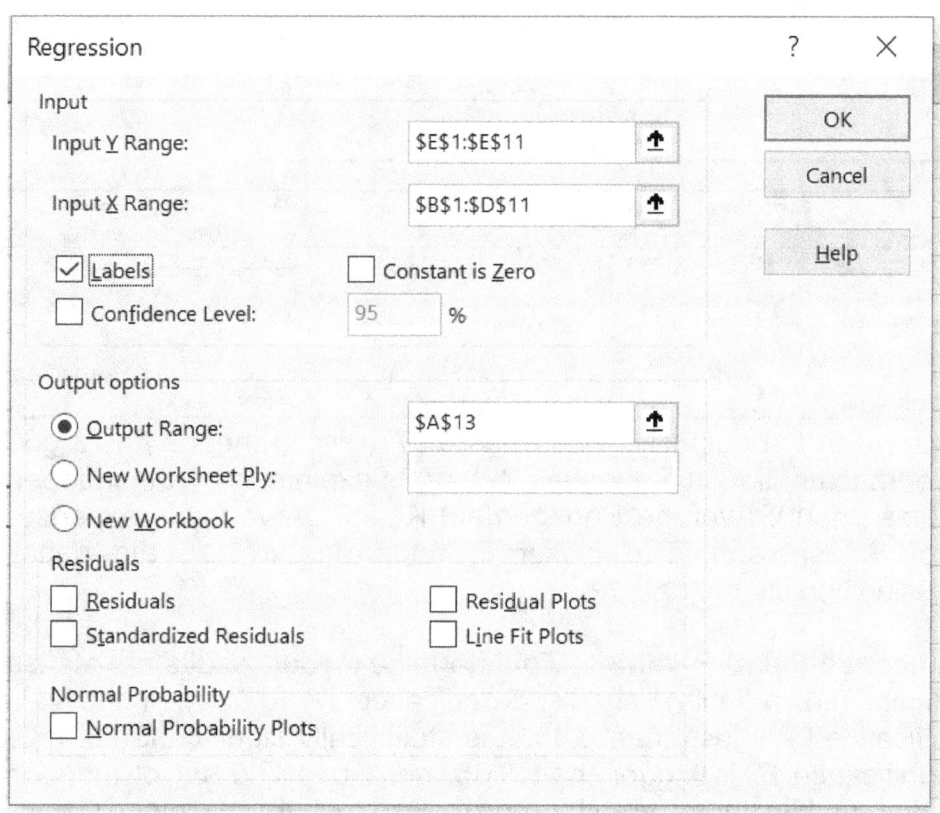

After you click "OK", Excel generates the following Summary Output. Round the numbers to 3 decimal places.

SUMMARY OUTPUT

Regression Statistics

Multiple R	0.988
R Square	0.977
Adjusted R Square	0.965
Standard Error	35.352
Observations	10

ANOVA

	df	SS	MS	F	Significance F
Regression	3	317141.486	105713.829	84.588	0.000
Residual	6	7498.514	1249.752		
Total	9	324640			

	Coefficients	Standard Error	t Stat	P-value	Lower 95%	Upper 95%	Lower 95.0%	Upper 95.0%
Intercept	335.899	194.523	1.727	0.135	-140.083	811.880	-140.083	811.880
Average Engagement Rating	41.958	13.123	3.197	0.019	9.847	74.069	9.847	74.069
Advertising Expense	4.818	3.277	1.470	0.192	-3.200	12.836	-3.200	12.836
No. of Salesperson	23.389	8.239	2.839	0.030	3.228	43.551	3.228	43.551

R Square: In the output, R Square is 0.977, which means it is a good fit. 97% of the variation in Revenue (Output) is explained by the independent variables (Input): Average Engagement Rating, Advertising expense, and No. of Salesperson. The closer R Square is to "1", the better the regression line fits the data.

Significance F and P-values: To determine if your results are statistically significant (i.e. reliable), check "Significance F" (0.001). If the value of "Significance F" is less than 0.05, it is statistically significant (i.e. reliable). If "Significance F" is bigger than 0.05, don't use this set of independent variables. Delete those variables with "P-value" that is bigger than 0.05 and run the regression again until "Significance F" drops below 0.05. Most or all your P-values should be lower than 0.05. In our example, "P-value" is 0.135, 0.019, 0.192 and 0.030, for Intercept, Average Engagement Rating, Advertising expense, and No. of Salesperson respectively.

Coefficients

From the Summary Output, the regression line is:

SUMMARY OUTPUT								
Regression Statistics								
Multiple R	0.988							
R Square	0.977							
Adjusted R Square	0.965							
Standard Error	35.352							
Observations	10							

ANOVA								
	df	SS	MS	F	Significance F			
Regression	3	317141.486	105713.829	84.588	0.000			
Residual	6	7498.514	1249.752					
Total	9	324640						

	Coefficients	Standard Error	t Stat	P-value	Lower 95%	Upper 95%	Lower 95.0%	Upper 95.0%
Intercept	335.899	194.523	1.727	0.135	-140.083	811.880	-140.083	811.880
Average Engagement Rating	41.958	13.123	3.197	0.019	9.847	74.069	9.847	74.069
Advertising Expense	4.818	3.277	1.470	0.192	-3.200	12.836	-3.200	12.836
No. of Salesperson	23.389	8.239	2.839	0.030	3.228	43.551	3.228	43.551

Revenue (Output)
= 335.899 + 41.958 * Average Engagement Rating + 4.818 * Advertising Expense + 23.389 * No. of Salesperson

Based on the above regression formula,
- For each unit increase in Average Engagement Score, Revenue increase by 41.958.
- For each unit increase in Advertising expense, Revenue increase by 4.818.
- For each unit increase in No. of Salesperson, Revenue increase by 23.389.

Coefficients can also be used for forecasting. For example, if "Average Engagement Rating" is 7, "Advertising Expense" is 80, "No. of Salesperson" is 3, then **predicted "Revenue"**
= 335.899 + 41.958 * Average Engagement Rating + 4.818 * Advertising Expense + 23.389 * No. of Salesperson
= 335.899 + (41.958 * 7) + (4.818 * 80) + (23.389 * 3)
= **$1085**

Reference
(1) Mohit Sharma, Talent Analytics: From Buzzword to Reality (2018), https://sightsinplus.com/2018/05/25/talent-analytics-from-buzzword-to-reality/ (24 September 2018)

2.3 Engagement Story 3: Engagement Storytelling & data visualisation

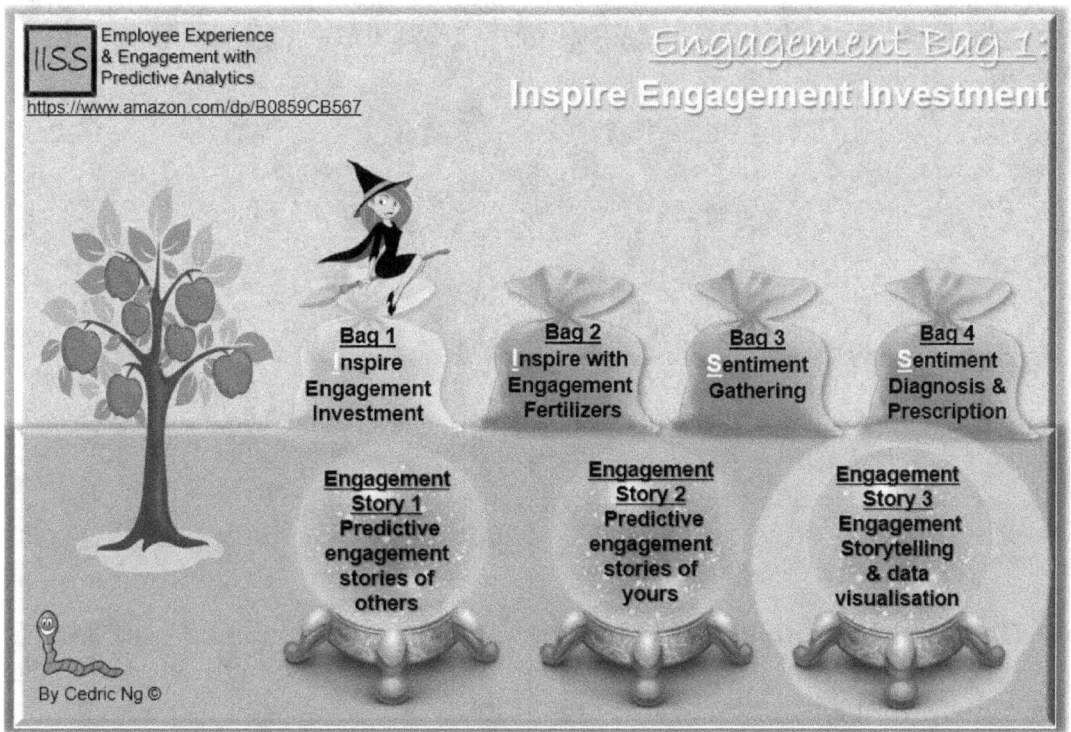

2.3.1 Data storytelling

According to Brent Dykes (2016), Data storytelling is a structured approach for sharing data insights, and it uses three key elements: **data**, **visuals**, and **narrative**.

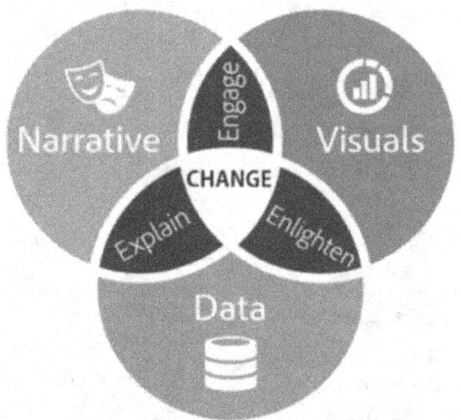

Source: https://www.forbes.com/sites/brentdykes/2016/03/31/data-storytelling-the-essential-data-science-skill-everyone-needs/#51c5d24752ad

- **Data** and **Narratives** explain: When narrative is combined with data, it helps to **explain** your data's context, source, relevance, what you did with it, what it is and why it is important.
- **Data** and **Visuals** enlighten. When visuals are combined with data, it can **enlighten** the audience to insights that wouldn't be visible without graphs or charts. Without data visualizations, interesting trends and data outliers would be hidden in the data tables rows and columns.
- **Narratives** and **Visuals** engage. When narrative and visuals are combined together, they **engage** the audience.

Data and **Narratives** and **Visuals** drive change. When you combine the right visuals and narrative with the right data, you have a data story that influences and drives **change**.

2.3.2 Stories & visuals that you can use to engage people

Stories have a lot of power — the power to inspire and change people. Stories can get inside people's minds and affect how they think, worry, and dream. Storytelling uses words and images to bring ideas and concepts to life. Storytelling can be used to illustrate the importance of an initiative, explain a product value, and engage people. Use inspiring workplace stories and quotes to engage your people by:
- Adding an inspiring story, quote or picture in your organization's newsletter
- Starting your meetings with an inspiring story, quote or picture.

Here are some stories & visuals that you can use to engage people:

We cannot force someone to hear a message they are not ready to receive, but we must never underestimate the power of planting a seed

Source:https://www.facebook.com/powerofpositivity/photos/a.162838037370/10156488056322371/?type=3&theater

Stories & Visuals: Diligence

To be sitting and doing nothing, you must be sitting very high up

A crow was sitting on a tree, doing nothing all day.

A small rabbit saw the crow, and asked him, "Can I also sit like you and do nothing all day long?"
The crow answered: "*Sure, why not.*"

So, the rabbit sat on the ground below the crow, and rested.

All of a sudden, a fox appeared,

Jumped on the rabbit... and ate it.

Source: https://image.slidesharecdn.com/corporatelessonsfromiimcalcuttabullshit-1228387565259380-8/95/corporate-lessons-from-iim-calcutta-bullshit-2-728.jpg?cb=1228359291

Stories & Visuals: Encouragement

I thought about quitting, but then I notice who was watching.

Source: https://i.pinimg.com/originals/5d/35/72/5d3572bd4af22d08e4b2679774200e0c.jpg

Remember that not getting what your want is sometimes a wonderful stroke of luck

Source:https://i1.wp.com/www.lifeadvancer.com/wp-content/uploads/2018/02/484-Remember-that-sometimes-not-getting-what-you-want.jpg?resize=731%2C466&ssl=1

Mr Bean got rejected by many TV shows due to his speaking disorder and stammering problem. Then he started his own show "Mr Bean" which became a global success.

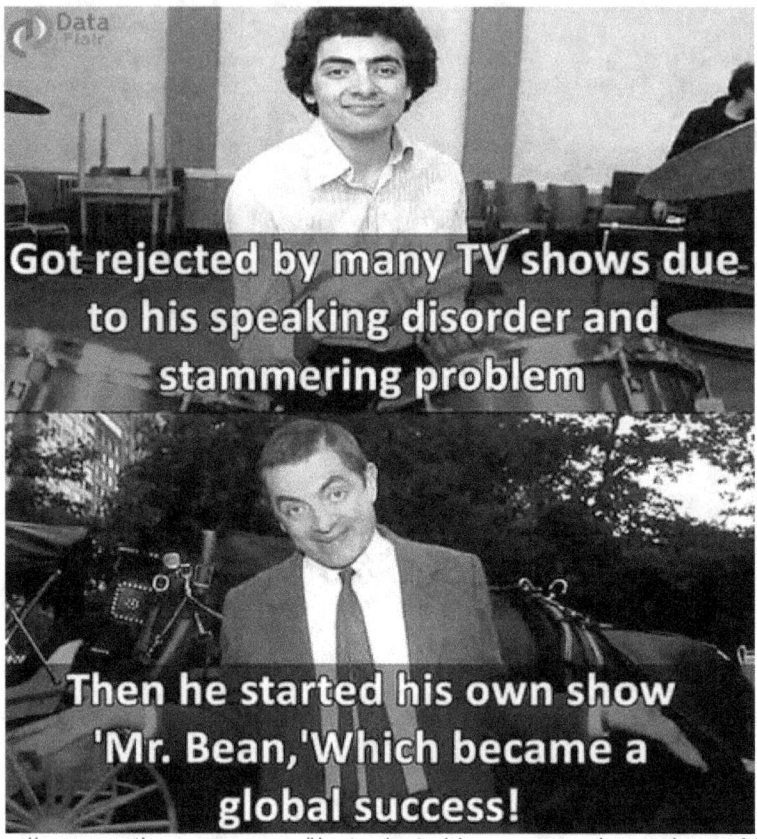

Source: https://conservativememes.com/i/got-rejected-by-many-tv-shows-due-to-his-speaking-20167935

Stories & Visuals: Office Gossip

Sometimes it's your mouth that is stopping you from moving up the career ladder.

Source: https://www.boredpanda.com/crocodile-life-animals-illustrations-keigo-japan/?utm_source=r.search.yahoo&utm_medium=referral&utm_campaign=organic

Stories & Visuals: Patience

Always let your boss have the first say.

A sales rep, an administration clerk and the manager are walking to lunch when they find an antique oil lamp. They rub it and a Genie comes out in a puff of smoke. The Genie says, "I usually only grant three wishes, so I'll give each of you just one." "Me first! Me first!" says the admin clerk "I want to be in the Bahamas, driving a speedboat, without a care in the world." Poof! She's gone. In astonishment, "Me next! Me next!" says the sales rep. "I want to be in Hawaii, relaxing on the beach with my personal masseuse, an endless supply of pina coladas and the love of my life." Poof! He's gone. "OK, you're up," the Genie says to the manager. The manager says, "I want those two back in the office after lunch."

Source: https://funnypictures.me/genie-grants-wishes/#

The man who moves a mountain begins by carrying away small stones

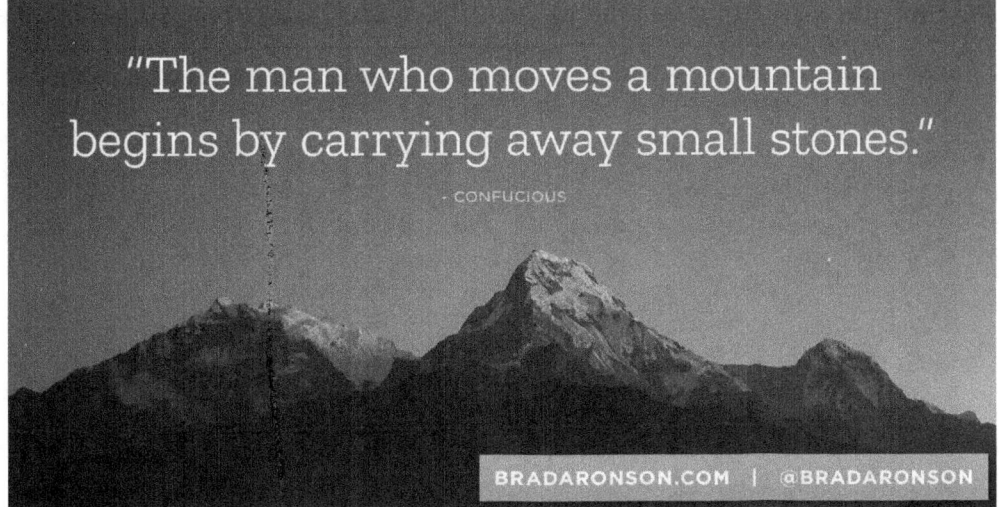

Source: https://www.bradaronson.com/wp-content/uploads/2019/10/Move-a-Mountain-life-quotes.jpg

Stories & Visuals: Risk taking

People who don't take risk generally make about two mistakes a year. People who do take risk generally make about two big mistakes a year.

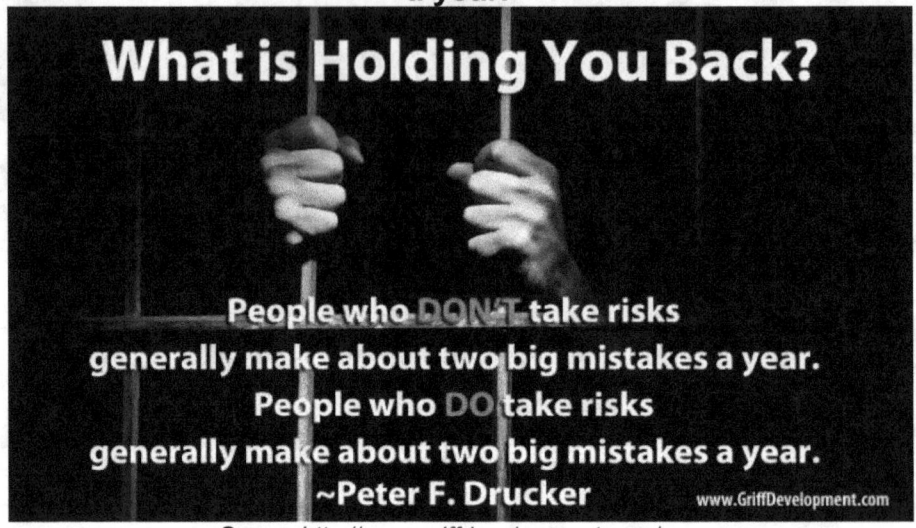

Source:http://www.griffdevelopment.com/wp-content/uploads/2017/05/AAEAAQAAAAAAANZAAAAJDcxZjk2NGJiLWYyYmUtNDQ1YS05MmFiLTExMGQ0NjJmYzBmYQ.jpg

Stories & Visuals: Teamwork

If you light a lamp for someone else, it will also brighten your path

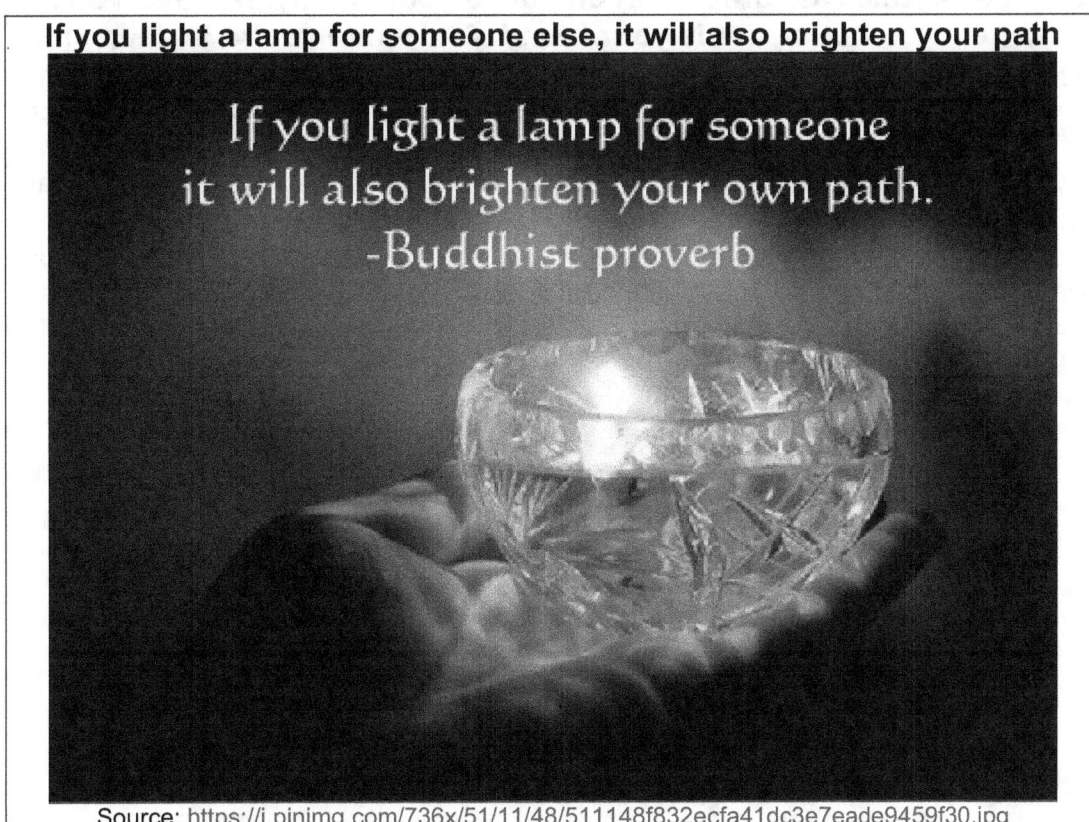

Source: https://i.pinimg.com/736x/51/11/48/511148f832ecfa41dc3e7eade9459f30.jpg

Not everyone who drops shit on you is your enemy. Not everyone who gets you out of shit is your friend. And when you're in deep shit, keep your mouth shut!

A little bird was flying south for the winter. It was so cold that the bird froze and fell to the ground in a large field. While it was lying there, a cow came by and dropped a load of hot, steaming dung on it. As the frozen bird lay there in the pile of shit, it began to realize how warm it was. The dung was actually thawing him out! He lay there all warm and happy and soon began to sing for joy. A passing cat heard the bird singing and came to investigate. Following the sound, the cat discovered the bird under the pile of cow dung. The cat promptly dug the bird out, killed him and ate him.

Source: https://1.bp.blogspot.com/-O8RwTfRH6AQ/W_9MdNumWHI/AAAAAAABG7Q/dv_ZErbfu84-gG8suaXtaM5LYaz1QR0HgCLcBGAs/s1600/frozen%2Bbird.JPG

Stories & Visuals: Work-life balance

2.3.3 Develop engaging stories

Different people can interpret the same data differently because data don't provide contextual information. Thus, you can't just show your data, you need to tell a story with it. Data storytelling connects the data, analysis, recommendations and visuals with the audience.

Steve Jobs used a three-act structure, with heroes, villains, and victims for his narrative: [1]

Act 1: The Setup — Why should I care?

Act 2: The Confrontation — How will your idea make my life better?

Act 3: The Resolution — What action do I need to take?

Act 1) The Setup - Why should I care?

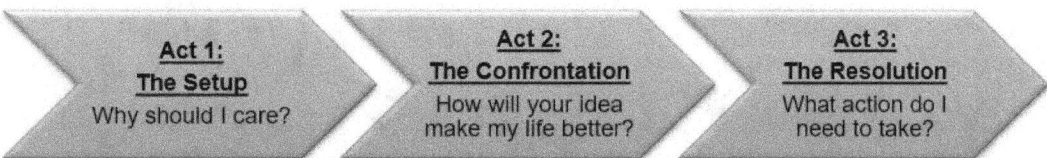

At the beginning of your story, you need to hook your audience to get them interested. Tell you audience what they'll get out from the presentation and why should they care. The setup introduces the characters (Heros, Villains, Victims, Supporting characters) in the context of a personal story.

- **Villains** are the **problems** that you are trying to overcome. For example, high staff turnover, absenteeism, performance issues, etc.
- **Heroes** are the **solutions** to the problems. For example, introducing incentive programmes, running training programmes, etc. You get to cast your organization, your team, yourself, or your project as the hero of your story.
- **Victims** are those harmed by the villains.
- **Supporting characters** are your data, not main character. Start your data storytelling with people, then data. Positioning your data story with the people you're addressing helps to make it more relatable to your audience.:
    - **Engagement**: Rather than saying, "Our company's average engagement score is seventy percent", say "two out of three of the people here are engaged."
    - **Attrition**: Rather than just saying, "Attrition rate is high", say "One-third of our new hires in Sales leave the company within 6 months, making it tough for us to achieve our sales targets.
    - **Goal setting**: Rather than saying, "We want to achieve 100% completed goal-setting," say "We want all our people here to be successful, thus we want all our manager here to tell our people what is expected of them at work through goal setting.

Act 2) The Confrontation - <u>How</u> will your idea make my life better?

| Act 1: The Setup — Why should I care? | Act 2: The Confrontation — How will your idea make my life better? | Act 3: The Resolution — What action do I need to take? |

The middle details the problem (villain) and eventually, the solution (hero) and supporting points go here. Here is where your show your analysis, findings, and the supporting facts, and answer how your solution will make the audience's life better.

- **Communicate your insights in a visual format**, instead of complicated correlation and regression tables. Visualization helps audiences to can grasp difficult concepts and identify patterns.
- **Avoid showing raw data or analysis**. If you present your data and analysis without sharing with the audiences your insights or interpretation, your audiences may draw their own conclusions to reinforce their preconceptions.
- **Select the most critical insights to share with your audience**. Don't overwhelm your audiences with the large number of insights that you generated. Examples of Apple's' memorable headlines are:
    - "iPhone - Apple reinvents the phone"
    - "MacBook Air - The world's thinnest notebook".
    - "iPOD - 1,000 songs in your pocket"

Act 3) The Resolution - What action do I need to take?

Act 1: The Setup	Act 2: The Confrontation	Act 3: The Resolution
Why should I care?	How will your idea make my life better?	What action do I need to take?

The ending is where you provide a road map to a better future, and leave the audience with a call to action. The Resolution has three steps:

- **Tell the Recommendations:** Write each recommendation clearly as a statement: Our insight demonstrated that if X is changed … Y will change… thus we recommend…
- **Encourage Action:** Tell your audience what you'd like them to do. Make it easy for your audience to act by breaking a big project into smaller milestones.
- **Get approval and implement it:** After telling the story, get approval from the sponsors and implement it. Evaluate your project along the way, analyzing the effectiveness of the implemented recommendations. In Analytics, we measure ourselves on the impact we made, not on the number of reports we produce, or the number of projects we completed. After you have made your recommendations, ask your sponsor whether they agree with your recommendations, and who will be responsible for implementing the recommendations?

References:
(1) Nick Mannon (2018) Persuasive Storytelling with Data Visualization. https://www.blastam.com/blog/persuasive-storytelling-with-data-visualization (25 November 2018)

2.3.4 Develop engaging visuals

Data visualization is the presentation of data in graphical or pictorial format, so that it is easier for audiences to see patterns or understand complex concepts. Because of how our brain processes information, it is easier to tell a story using pictures or graphs to visualize complex data than poring over columns and rows of Excel data.

Consider color groups and contrast

It is important to consider color groups, when combining colors for your presentation. Colors are classified into broad groups: [1]

- **Warm colors (red, orange, yellow):** Warm colors pop out and attract attention, especially bright red. It is <u>ok to combine warm colors with each other</u> and shades of brown.
- **Cool colors (green, blue, purple):** Cool colors recede in the background and don't draw attention, especially darker shades. It is <u>ok to combine cool colors with each other</u> and shades of gray.
- **Neutral colors (white, black, beige):** Neutral colors goes well with all colors. For professional presentations, use white or light beige on a dark background. Else, use black or dark color on a light background.

Avoid combining colors across the warm/cool boundary as they cause eye strain, especially for bright blue + red combination, and red + green combination. An issue with the red + green combination is that 7 percent of men and 1 percent of women have cannot differentiate red and green colors because it is a common type of color blindness. [1]

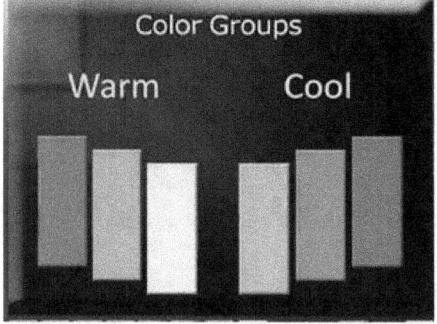

Source: https://support.office.com/en-us/article/combining-colors-in-powerpoint---mistakes-to-avoid-555e1689-85a7-4b2e-aa89-db5270528852

Differentiate by color rather than shape

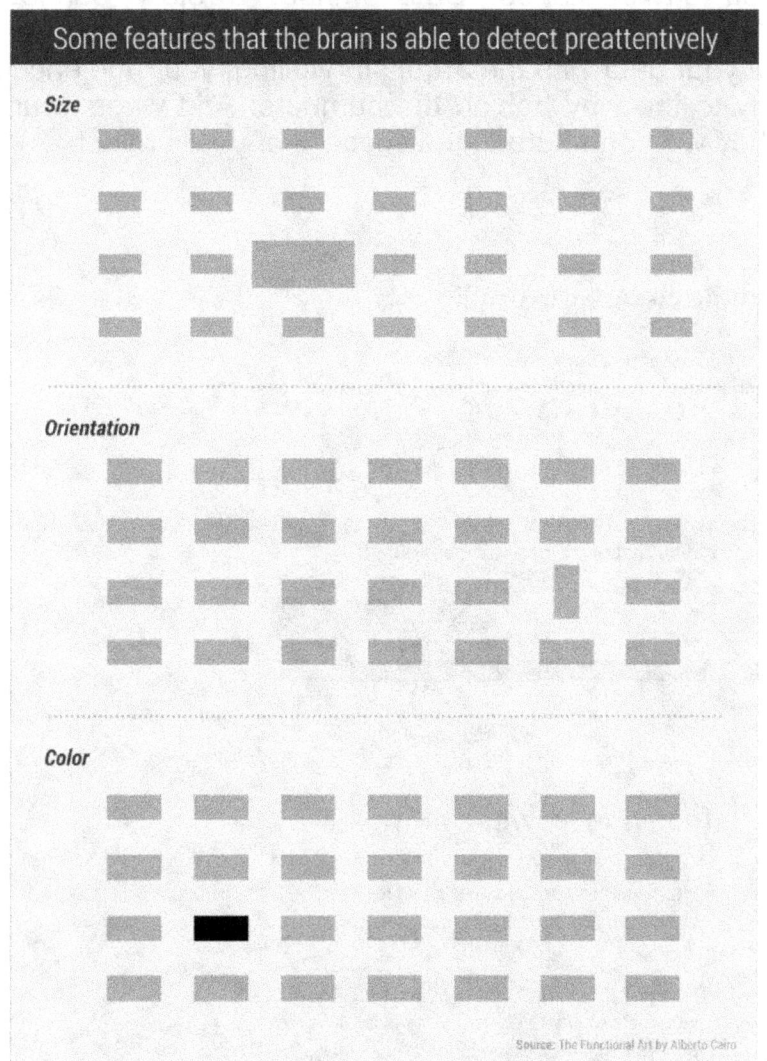

Source: https://www.crazyegg.com/blog/data-storytelling-5-steps-charts/

Fewer Bullet Points, More Narrative

Our hearts and minds crave stories. Data without a story cannot emotionally connect and engage your data with your audience. By combining the power of your data with the art of storytelling, your audience will connect emotionally to the why behind the numbers. And when your people are engaged, they will contribute to the success of your project. [2]

Avoid this,

Dogs Have an Incredible Sense of Smell

- A dog's sense of smell is much keener and stronger than a human's.
- Their noses are about a million times more sensitive than ours.
- Their noses are so powerful that they can even smell disease.
- Research has found that dogs have an incredible ability to detect the smell of a range of organic compounds that indicate the human body isn't working properly.
- There are now diabetic alert dogs who signal the human when they pick up the distinct scent that is released when a human's insulin levels drop.

Use this,

Lifesaving Power of Smell

Marie has Type 1 Diabetes. Her service pup, Fitz, uses his keen sense of smell to detect and signal Marie when her insulin levels drop.

Research has found that dogs have an incredible ability to detect the smell of a range of organic compounds that indicate when the human body isn't working properly.

Source: https://patronmanager.com/blog-power-up-your-powerpoint-with-storytelling/

Less Text, More Visuals

Avoid putting blocks of text on your PowerPoint presentation. Use pictures or diagrams with keywords and short phrases. Let your speech to attract your audience.

Avoid this,

Use this,

Source: https://patronmanager.com/blog-power-up-your-powerpoint-with-storytelling/

Minimize Stand Alone Data

Don't just present a chart on its own. Present the figures with a story — they create connection and action.

Avoid this,

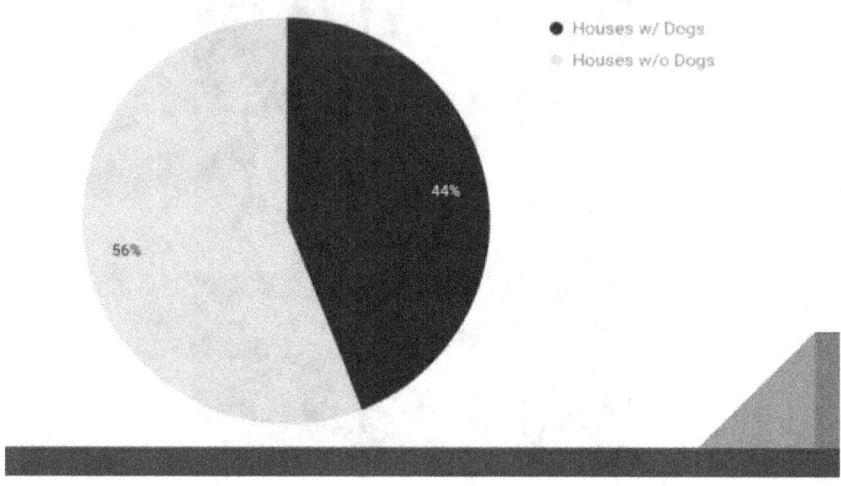

Use this,

Source: https://patronmanager.com/blog-power-up-your-powerpoint-with-storytelling

Declutter Your Chart

Declutter your visualizations by removing unnecessary and repetitive information. Summarize your information if your audience don't need to know the details. Ask yourself:
- Do I need these details to convey my message across?
- Can I summarize this information?"

Here are tips to declutter your visualizations:
- Avoid using more than 3 colors.
- Use grey color for items in the background that are not so important.
- Label columns, lines or segments directly instead of using a legend, so that your audience eyes do not need to move all over your slide.
- Remove chart borders and gridlines.

Avoid this,

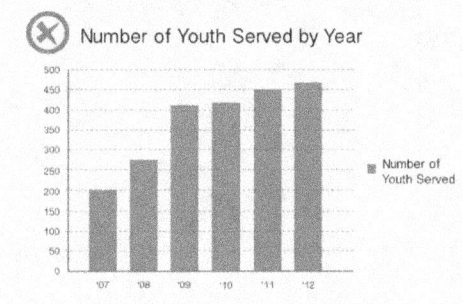

Use this,

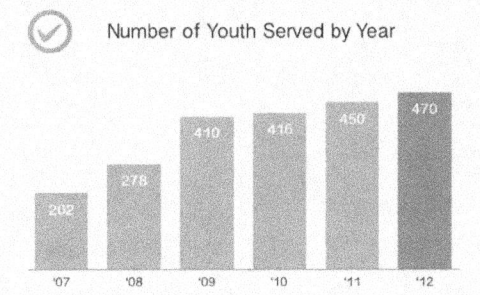

Source: https://www.crazyegg.com/blog/data-storytelling-5-steps-charts/

Push Secondary Information to the Back

To focus your audience's attention, highlight the most important items by using a different color and thicker line. Push all the secondary information to the background by using light gray color. (5)

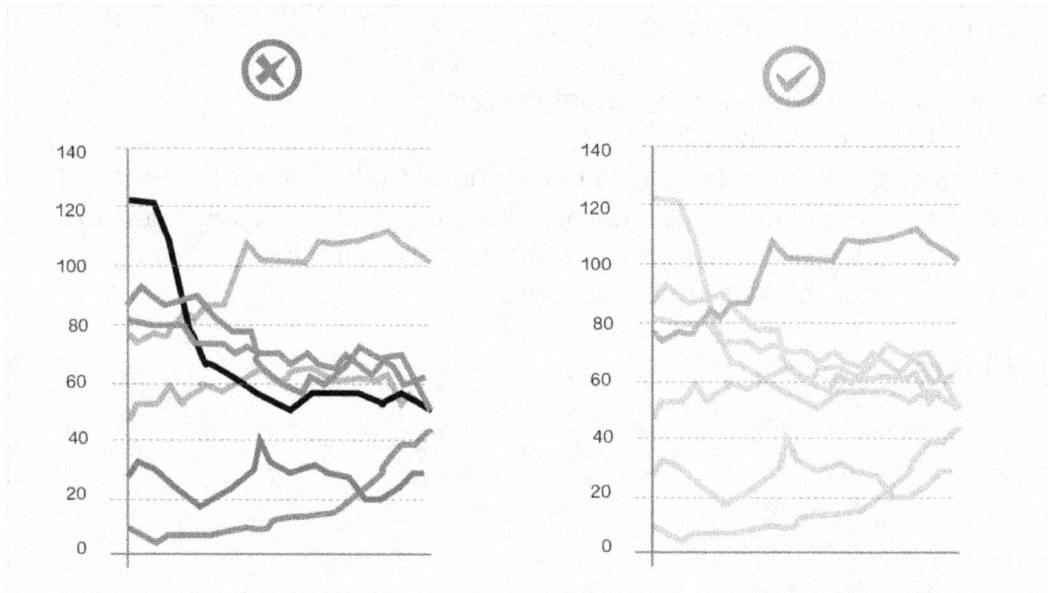

Source: https://www.crazyegg.com/blog/data-storytelling-5-steps-charts/

Push Secondary Information to the Back

In the graph below, the message is: (5)

> Since mid-2014, the supply of oil has exceeded demand, resulting in excess (blue bars).

However, the main message was distracted because too many colors was used, and secondary information were in bold.

Avoid this,

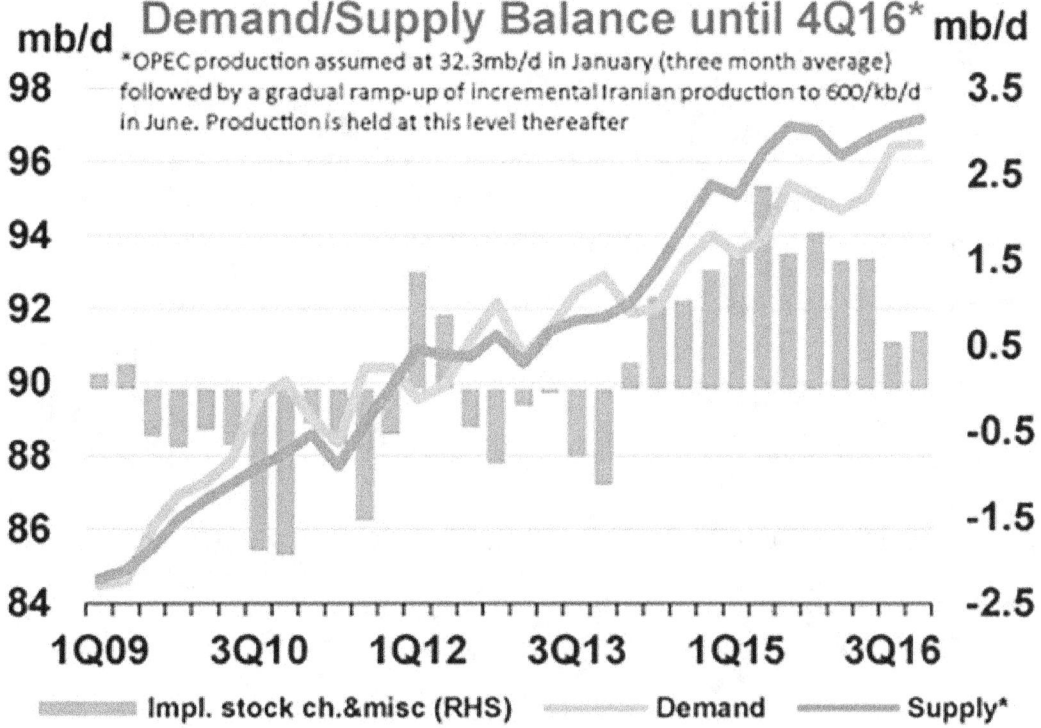

Source: https://www.crazyegg.com/blog/data-storytelling-5-steps-charts/

Below is an improved version of the above graph. To focus your audience's attention, the most important items (change in supply and demand starting in mid-2014) were highlighted by using a different color. All the secondary information was push to the background by using light gray color, and the axis values, title and subtitle are deemphasized. (5)

Use this,

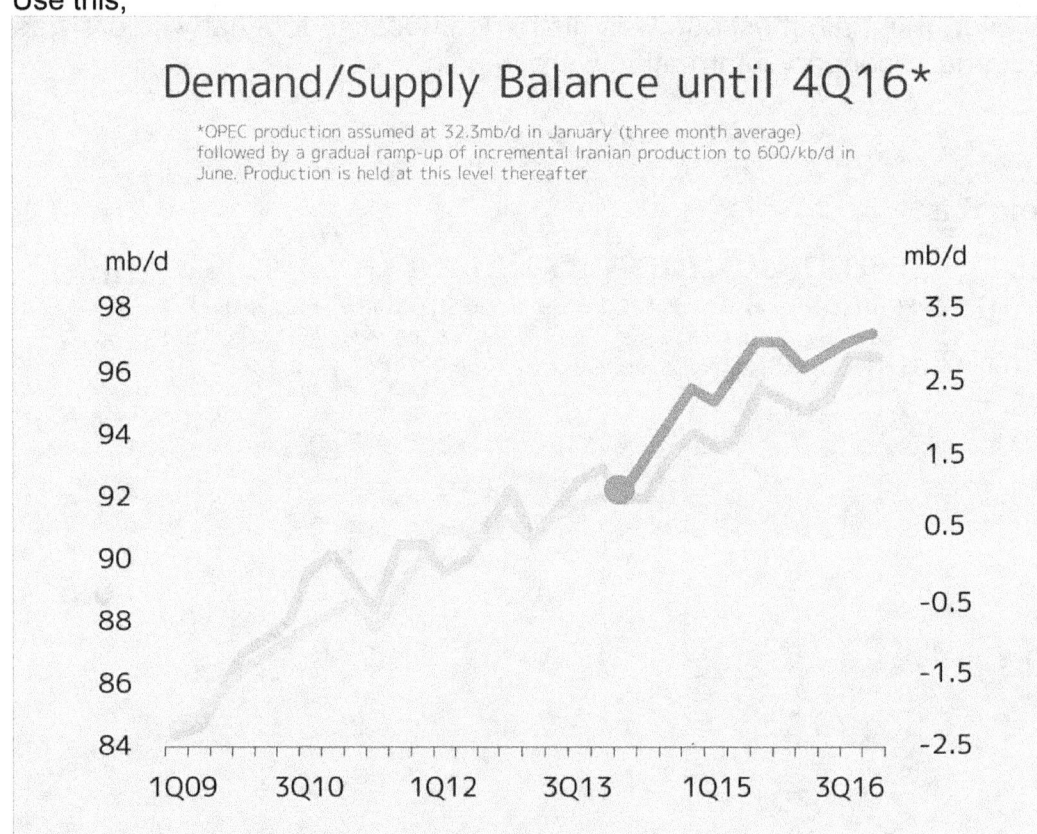

Source: https://www.crazyegg.com/blog/data-storytelling-5-steps-charts/

Use Slope-graphs to Compare Rate of Change

When you want to compare the rate of change between two points in time, it is easier to visualize through the slope of the line, than through bar charts. (5)

Avoid this,

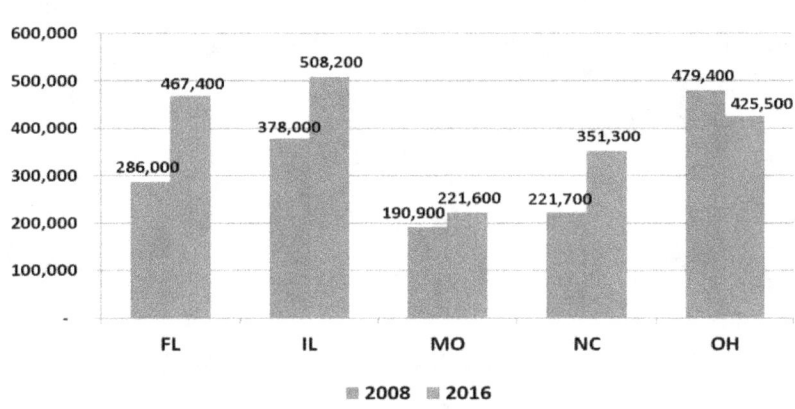

Use this,

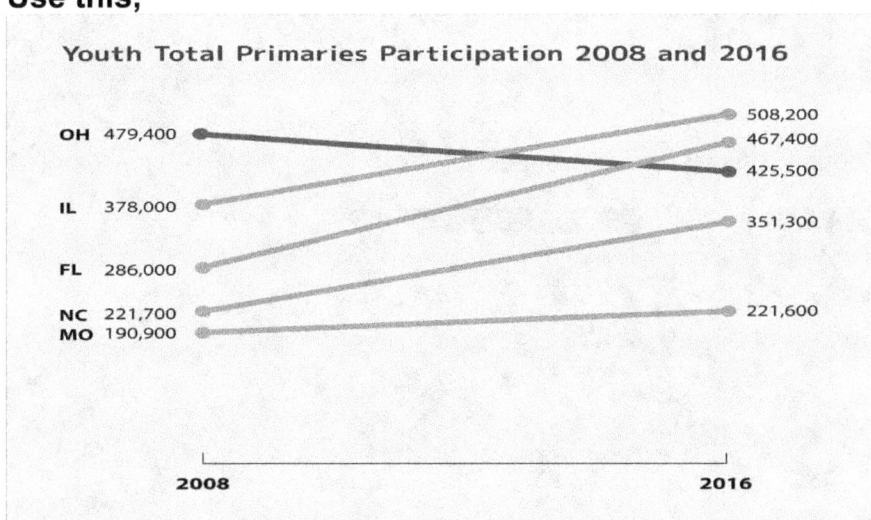

Source: https://www.crazyegg.com/blog/data-storytelling-5-steps-charts/

Avoid using legends

It is tedious to read the chart below because the bubbles are not proportional to each other. The reader needs to look back and forth between the bubbles and the corresponding legend. Side-by-side bar charts is a better way to present this information because they all start at the same baseline. The difference between values can be easily compared by just looking at the length of the bars, instead of a legend. (5)

Avoid this,

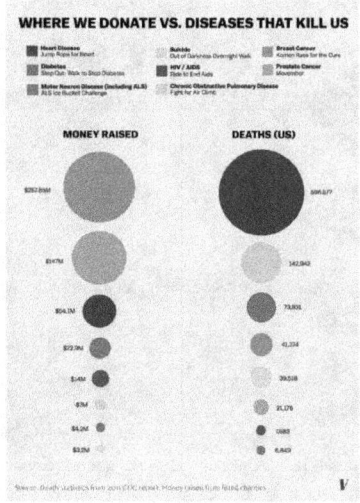

Use this,

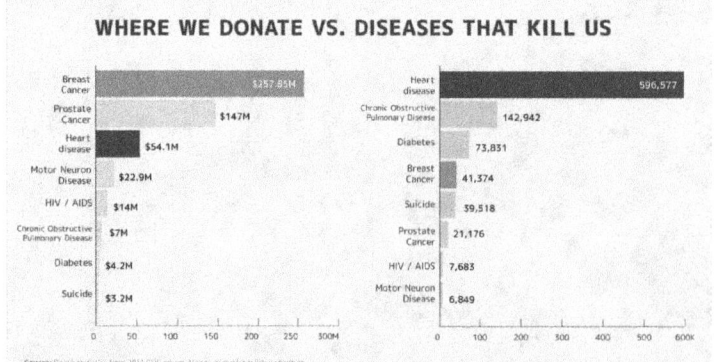

Source: https://www.crazyegg.com/blog/data-storytelling-5-steps-charts/

References:
(1) Robert Lane (2018) Combining colors in PowerPoint – Mistakes to avoid. https://support.office.com/en-us/article/combining-colors-in-powerpoint---mistakes-to-avoid-555e1689-85a7-4b2e-aa89-db5270528852 (23 November 2018)
(2) Rebekah Pearson (2018) https://patronmanager.com/blog-power-up-your-powerpoint-with-storytelling/ (23 November 2018)
(5) Nayomi Chibana (2017) https://www.crazyegg.com/blog/data-storytelling-5-steps-charts/ (23 November 2018)

3.0 Engagement Bag 2: Inspire with Engagement Fertilizers

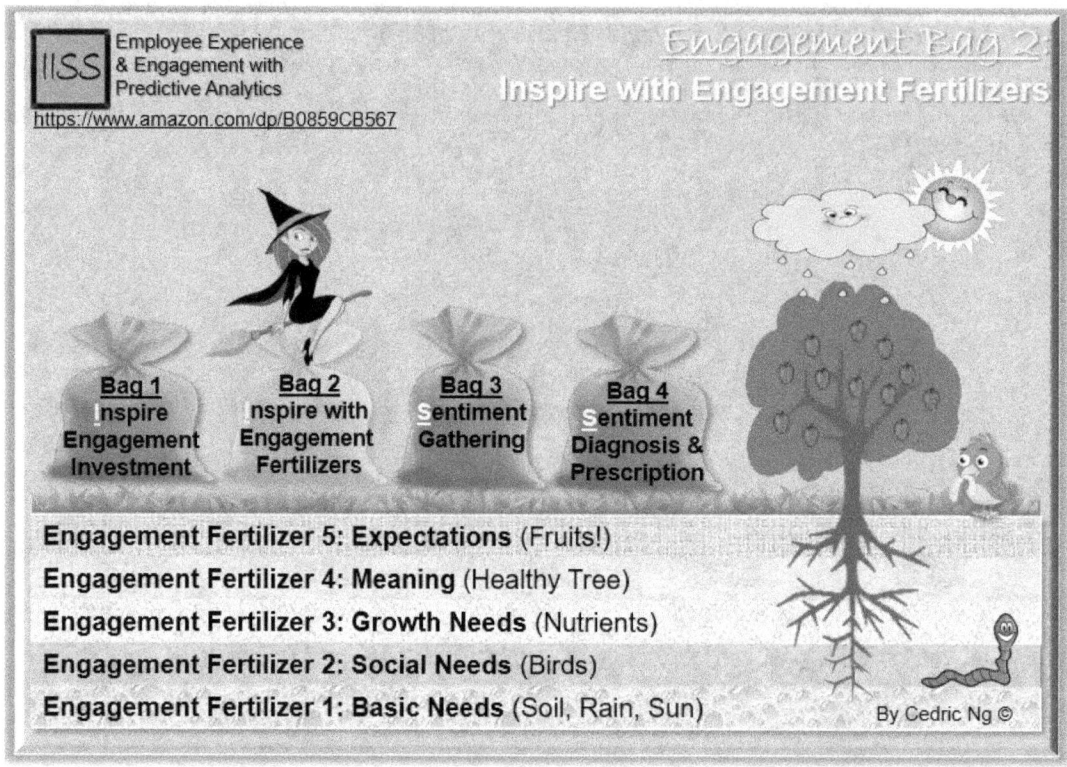

In IISS, "**Inspire with Engagement Fertilizers**" is the second bag, because you can only successfully implement engagement programs, after you get your organization and managers support. Making employees happy, doesn't mean they will work hard for the organization. In IISS, "**5 Engagement Fertilizers**" are used to create the fertile soil for great employee experience and engagement.

- **Engagement Fertilizer 1: Basic Needs** - Soil, Rain, Sun
- **Engagement Fertilizer 2: Social Needs** - Birds
- **Engagement Fertilizer 3: Developmental Needs** - Nutrients
- **Engagement Fertilizer 4: Meaning** - Healthy Tree
- **Engagement Fertilizer 5: Expectations** - Fruits!

Although the 5 "Engagement Fertilizers" are described as hierarchical, application of the Fertilizers need not be rigid. Without fertilizer 1 (soil, rain, sun), you can't grow satisfaction. But to grow engagement, you need the rest of the fertilizers (right amount of rain, sun, nutrients, and the help of birds).

3.1 Engagement & Motivation Theories

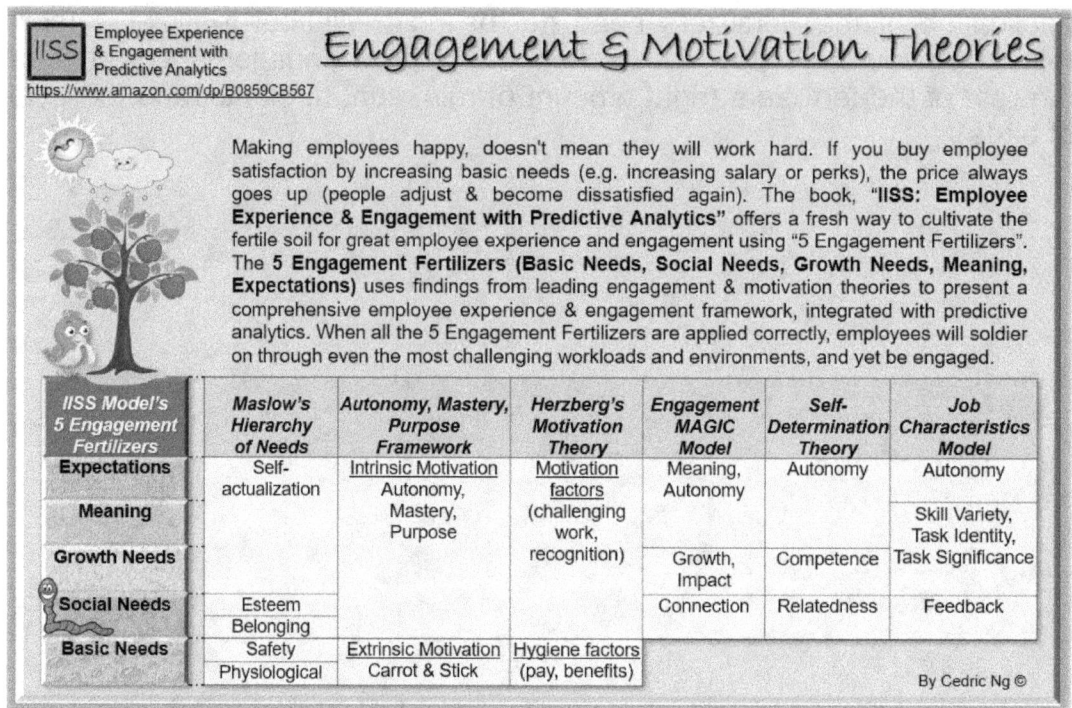

Making employees happy, doesn't mean they will work hard for you. If you buy employee satisfaction by increasing basic needs (e.g. increasing salary or perks), the price always goes up (people adjust & become dissatisfied again). IISS offers a fresh way to cultivate the fertile soil for great employee experience and engagement using "5 Engagement Fertilizers". The IISS model & it's **5 Engagement Fertilizers** (Basic Needs, Social Needs, Growth Needs, Meaning, Expectations) is aligned with leading engagement & motivation theories, & it uses findings from leading theories & research to present a comprehensive employee experience & engagement framework, integrated with predictive analytics. When all the 5 Engagement Fertilizers are applied correctly, employees will soldier on through even the most challenging workloads and environments, and yet be engaged.

To be an employee engagement subject matter expert, it is useful for HR practitioners to know the leading employee engagement and motivation theories that can be applied in different situations.:
- Maslow's Hierarchy of Needs Theory
- Herzberg's Motivation Theory
- Self-Determination Theory
- Autonomy, Mastery, Purpose framework
- Engagement MAGIC Model
- Job Characteristics Model

3.1.1 Maslow's Hierarchy of Needs

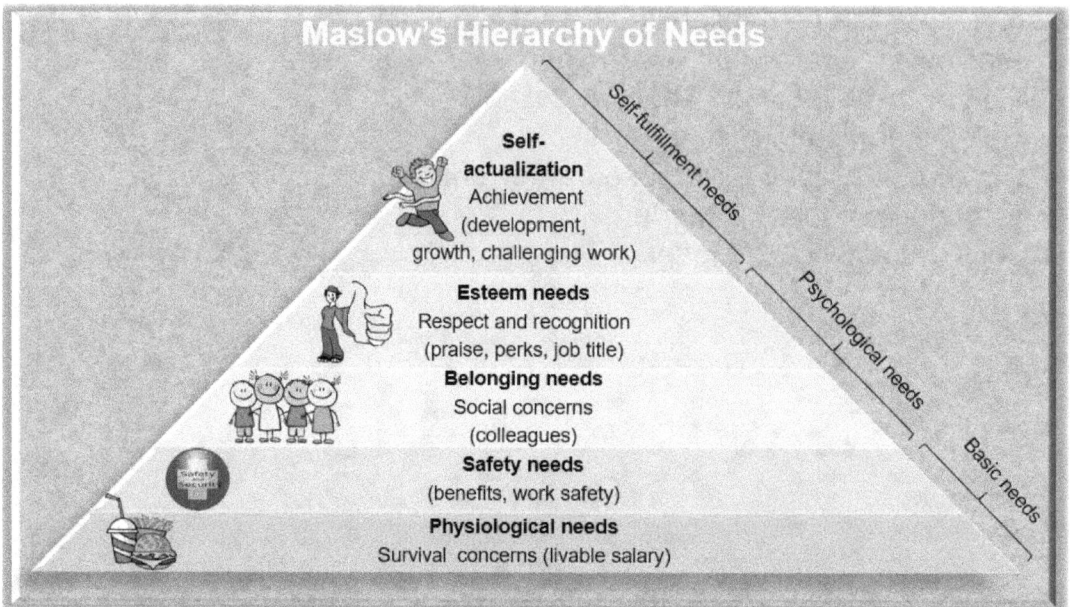

What motivates people? Maslow's hierarchy of needs is one of the best-known theories of motivation. The fields of education and business have been particularly influenced by the theory. Maslow first introduced his concept of a hierarchy of needs in his 1943 paper "A Theory of Human Motivation" and his subsequent book "Motivation and Personality". This hierarchy suggests that people are motivated to fulfill basic needs before moving on to other, more advanced needs. While some of schools of thought at the time tended to focus on problematic behaviors, Maslow was more interested in learning about what makes people happy and the things that they do to achieve that aim. Maslow believed that people have an inborn desire to be self-actualized, that is, to be all they can be. In order to achieve these ultimate goals, a number of more basic needs must be met such as the need for food, safety, belonging, and self-esteem. There are five different levels of Maslow's hierarchy of needs. Maslow's hierarchy is most often displayed as a pyramid. The lowest levels of the pyramid are made up of the most basic needs, while the most complex needs are at the top of the pyramid. [1]

1. **Physiological Needs**. Needs at the bottom of the pyramid are basic physical requirements that people need to survive (e.g. food, water). In the workplace, this refers to survival issues (e.g. livable salary) and workplace needs like comfortable work environment, snacks, coffee, etc. People need to deal with the survival needs before other needs. If they do not have basic food, water and shelter, they are unlikely to be concerned about learning new skills. Once these lower-level needs have been met, people can move on to the next level of needs.

2. **Safety Needs.** As we move up to the second level of Maslow's hierarchy of needs, the requirements become more complex. At this level, the needs for safety become primary. Finding a job, health insurance, savings money, and moving into a safer neighborhood are all examples of actions motivated by the safety needs. Safety needs encompass physical safety and financial safety.
   - **Physical safety** needs include health and wellness, and being safe from viruses, violence and natural disasters.
   - **Financial safety** needs involve having stable job, being able to pay bills, and being covered with life insurance. If employees have to worry about their personal safety or financial security, they may become disengaged.

3. **Belonging Needs.** Once Physiological and Safety needs are fulfilled, belonging needs begin to take greater precedence. Belonging Needs have to do with people's desire to be part of a group, or a need to form and maintain lasting social connections. Gallup includes the following question in its employee engagement survey: "Do you have a best friend at work?" This question matters because based on Gallup's research, employees who answer affirmatively are more likely to be engaged.

4. **Esteem Needs**. After the belonging needs have been addressed, the need to gain esteem and recognition becomes more important. People want to feel appreciated, respected, and recognized for their work and accomplishments. They want to feel that they are making important contributions to their department, organization and society.

5. **Self-Actualization Needs** is at the peak of the hierarchy. Self-actualization involves a person's need to fulfill their total potential and to become the best that they can possibly be. What this entails exactly depends upon what is important to the individual. In the workplace, take into account each team members personal goals when assigning tasks. Self-actualization is not a destination or end-point. Self-actualization is characterized by the need to continue becoming better. Self-actualizing people are self-aware, concerned with personal growth, and less concerned with the opinions of others.

<u>References</u>
(1) Kendra Cherry (2019) The 5 Levels of Maslow's Hierarchy of Needs <u>https://www.verywellmind.com/what-is-maslows-hierarchy-of-needs-4136760</u> *(13 February 2020)*

3.1.1.1 Deficiency Needs vs. Growth Needs

Deficiency needs - Physiological, Safety, Belonging and Esteem needs are deficiency needs, which arise due to deprivation. Satisfying these lower-level needs is important in order to avoid unpleasant feelings or consequences. [1]

- **Basic needs:** Physiological and Safety are often referred to as "basic needs".
- **Psychological needs:** Belonging and Esteem needs are often referred to as "psychological needs".

Growth needs - Maslow termed the highest level of the pyramid as growth needs. These needs don't stem from a lack of something, but rather from a desire to grow as a person. [1]

While Maslow's theory is generally portrayed as a fairly rigid hierarchy, Maslow noted that the order in which these needs are fulfilled does not always follow this standard progression. For example, he noted that for some individuals, the "self-esteem needs" is more important than the "belonging needs". For others, the need for creative fulfillment (Self-Actualization Needs) may supersede even the most basic "physiological needs". [1]

References
(1) Kendra Cherry (2019) The 5 Levels of Maslow's Hierarchy of Needs https://www.verywellmind.com/what-is-maslows-hierarchy-of-needs-4136760 (13 February 2020)

3.1.1.2 Why is Maslow's Hierarchy of Needs so Influential?

Focused on the development of healthy individuals
Maslow's hierarchy of needs represents part of an important shift in psychology. Rather than focusing on abnormal behavior and development, Maslow's humanistic psychology was focused on the development of healthy individuals. [1]

Human needs are universal across different cultures
In research published by Tay and Diener (2011), the researchers analyzed participants from more than 120 different countries, and found that there do appear to be human needs that are universal across different cultures. Their research also suggests that while these needs exist in cultures all over the world, they do not follow the order presented in Maslow's hierarchy. Instead, Tay and Diener suggest that the needs are dynamic and not independent from one another. Maslow's hierarchy may not follow the order in which it is usually presented, but his theory does offer a useful framework for understanding how different needs motivate human behavior. [2]

References
[1] Kendra Cherry (2019) The 5 Levels of Maslow's Hierarchy of Needs https://www.verywellmind.com/what-is-maslows-hierarchy-of-needs-4136760 (13 February 2020)
[2] Kendra Cherry (2017) Maslow's Hierarchy of Needs https://www.explorepsychology.com/maslows-hierarchy-of-needs/ (13 February 2020)

3.1.1.3 Critique of Maslow's theory

While popular, Maslow's theory has not been without criticism. Maslow's hierarchy of needs theory is lacking when it comes to empirical research and scientific support.

People's needs are not necessarily hierarchical
While research supports the idea that these needs are important, these needs may not necessarily follow a strict hierarchy and these needs may exist and interact in a dynamic and continually changing way. In a study published in 2011, researchers from the University of Illinois set out to put the hierarchy of needs to the test. They discovered that while the fulfillment of the needs was strongly correlated with happiness, people who struggle to fill their most basic physiological and security needs may still pursue esteem and self-actualization needs. Such results suggest that while these needs can be powerful motivators of human behavior, they do not necessarily take the hierarchical form that Maslow described. [1] A person can focus on more than one need. Promotion not only satisfies "esteem needs" and "self-actualization needs", but also a person's "physiological needs" (higher salary). Although the needs are described as hierarchical, application of the theory should not be rigid.

Difficult to test Maslow's theory
Maslow's definition of self-actualization is difficult to test scientifically. [1]

Biased sample
Despite the popularity of Maslow's hierarchy, there is not much recent data to support it. Maslow's sample was biased as it included only a small sample of people from a similar culture (his sample was famous white men living in the Western world). Thus, it is difficult to generalized his findings to women, minorities, or people from non-Western cultures.[2]

References
(1) Kendra Cherry (2019) *The 5 Levels of Maslow's Hierarchy of Needs* *https://www.verywellmind.com/what-is-maslows-hierarchy-of-needs-4136760* *(13 February 2020)*
(2) Kendra Cherry (2017) *Maslow's Hierarchy of Needs* *https://www.explorepsychology.com/maslows-hierarchy-of-needs/* *(13 February 2020)*

3.1.1.4 Engaging Employees using the Maslow hierarchy of needs

1. Learn what is important to your employees by listening, observing and asking.
2. Identify each employee's most pressing need (Physiological needs, Safety needs, Belonging needs, Esteem needs, Self-actualization)
3. Align rewards and your communications style to each employee's most pressing need

3.1.1.5 "Maslow's hierarchy of needs" vs "5 Engagement Fertilizers"

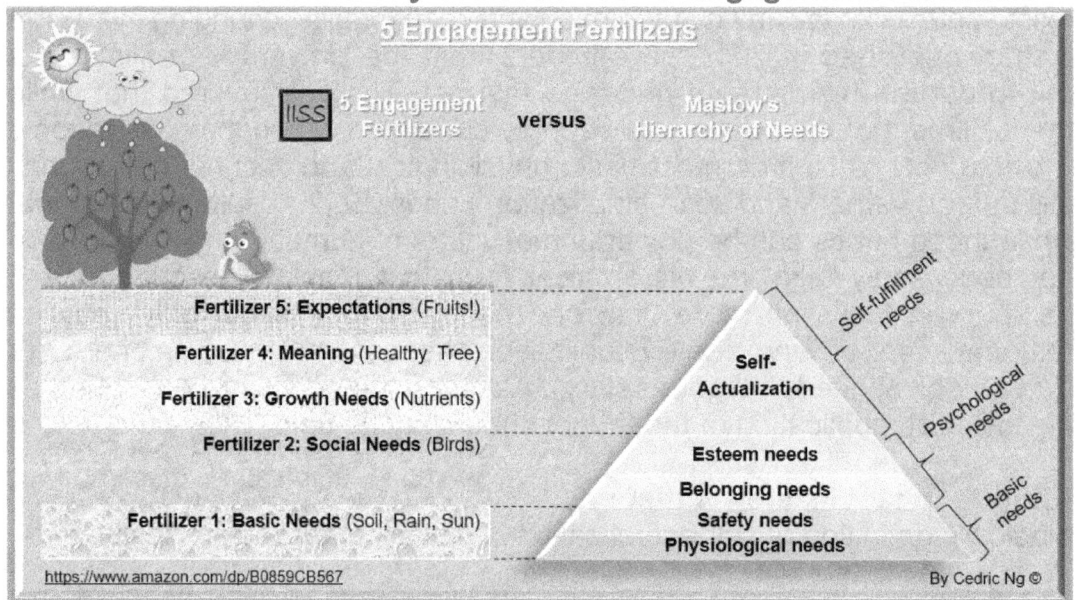

The IISS model's 5 Engagement Fertilizers is similar to Maslow's hierarchy of needs.:

- **Fertilizer 1 (Basic Needs)** is similar to Maslow's "Basic needs" and "Psychological needs".
- **Fertilizer 2 (Social Needs)** is similar to Maslow's "Belonging needs" and "Esteem needs".
- **Fertilizer 3 (Growth Needs), Fertilizer 4 (Meaning)**, and **Fertilizer 5 (Expectations)** is similar to Maslow's "self-actualization".

3.1.2 Herzberg's Motivation Theory

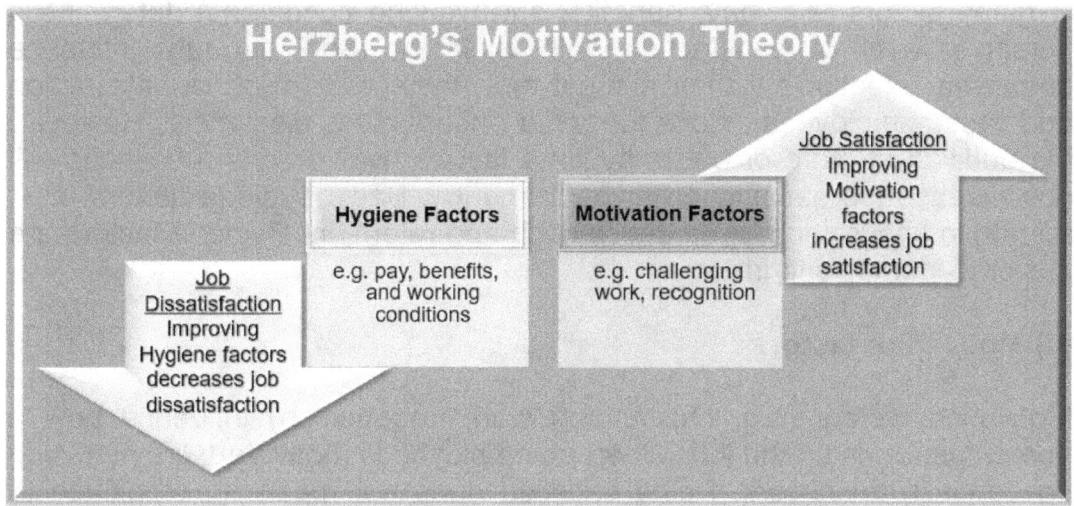

According to Frederick Herzberg's Motivation Theory (also called Herzberg's Motivation-Hygiene Theory, Herzberg's Two-Factor Theory), people are influenced by Hygiene (e.g. pay, benefits, and working conditions) and Motivation factors (e.g. challenging work, recognition). Both factors are key components of employee engagement.

(i) Hygiene factors

Hygiene factors (e.g. pay, benefits, and working conditions) determine a person's level of satisfaction with their job and strongly influence employee retention. If they are not met, they lead to job dissatisfaction and cause employees to look for better opportunities elsewhere. However, the addition of more or better hygiene factors over a certain baseline will not increase job satisfaction or performance. Basic hygiene factors must be met to ensure employee satisfaction and retention. Hygiene factors are easier to identify and improve.

(ii) Motivation factors

Motivation factors (e.g. challenging work, recognition) influence how a person performs on the job. When an employee is motivated, they strive to do better in their work. Being satisfied does not mean an employee to work harder. Also, an employee can be highly motivated but not satisfied with the job. They might find the work interesting and challenging, but if they think they can be paid more at a different company, they will not be satisfied. An employee must also feel motivated in order to perform at a high level. Specific motivation factors differ for each employee and are most influenced by the employee's supervisor.

3.1.2.1 Why is Herzberg's Motivation Theory so Influential?

Supported with empirical data
Herzberg's motivation theory emerged from a collection of data gathered by the interview of 203 accountants and engineers within the Pittsburgh area. The interview process consisted of asking the respondents to describe a work situation where they felt very happy as well as very unhappy. The analysis of the responses confirmed the proposed hypothesis, where some factors where contributors to job satisfaction, while others were not. [1]

Emphasize on motivation from within
Herzberg's motivation theory emphasizes on motivation coming from within the employees themselves rather than focusing on other external factors. This theory ensures that on the part of the company, it can improve its working environment and conditions so that employees feel motivated to work hard.

Compatible with Maslow's hierarchy of needs
Herzberg's framework seems compatible with Maslow's hierarchy of needs. Herzberg's Hygiene factors are roughly equivalent to Maslow's lower order needs, and Herzberg's motivational factors are somewhat equivalent to higher-order needs. [2]

References
(1) leadership-central (2020) Two Factor Theory   *https://www.leadership-central.com/two-factor-theory.html* (13 February 2020)
(2) Nidhisingh B (2020) Herzberg's Two Factor Theory of Motivation *http://www.economicsdiscussion.net/motivation/herzbergs-two-factor-theory-of-motivation/31821* (13 February 2020)

3.1.2.2 Critique of Herzberg's Motivation Theory

Job satisfaction may not result in higher productivity
The biggest disadvantage of this theory is that it assumes that job satisfaction equals higher productivity. Job satisfaction is one of the factors behind the increase in job productivity, but not the only factor. If organization is thinking that an increase in job satisfaction will lead to an increase in job productivity, then it may set itself for disappointment.

What motivates one individual might be a de-motivator for another individual.
Another problem with the Herzberg theory is that different people have different meaning when it comes to job satisfaction. As factors that motivate can change during a person's lifetime, younger employees may see job security as a hygiene factor, whereas older employees may see job security as a motivator.

Ignores External Factors
Another limitation of this theory is that it ignores external factors. If a competitor organization is paying a higher salary for the same job, its employees will not be satisfied even if the organization has implemented all factors of Herzberg theory.

Small sample size using white collar occupations only
Herzberg's theory is based on survey of 200 employees who were engineers or accountants. Thus, Herzberg's theory cannot be generalised because the sample size is too small and two white collar occupations cannot represent entire workforce. This theory is more applicable to white collar workers.[1]

Fails to link organizational goal needs and individual needs. [1]

Fails to attached importance to pay, status, relationships
Herzberg's theory does not attach much importance to pay, status, or interpersonal relationships which are important contents of satisfaction. [1]

References
(1) Nidhisingh B (2020) Herzberg's Two Factor Theory of Motivation
http://www.economicsdiscussion.net/motivation/herzbergs-two-factor-theory-of-motivation/31821
(13 February 2020)

3.1.2.3 "Herzberg's Motivation Theory" vs "IISS Model's Engagement Fertilizers"

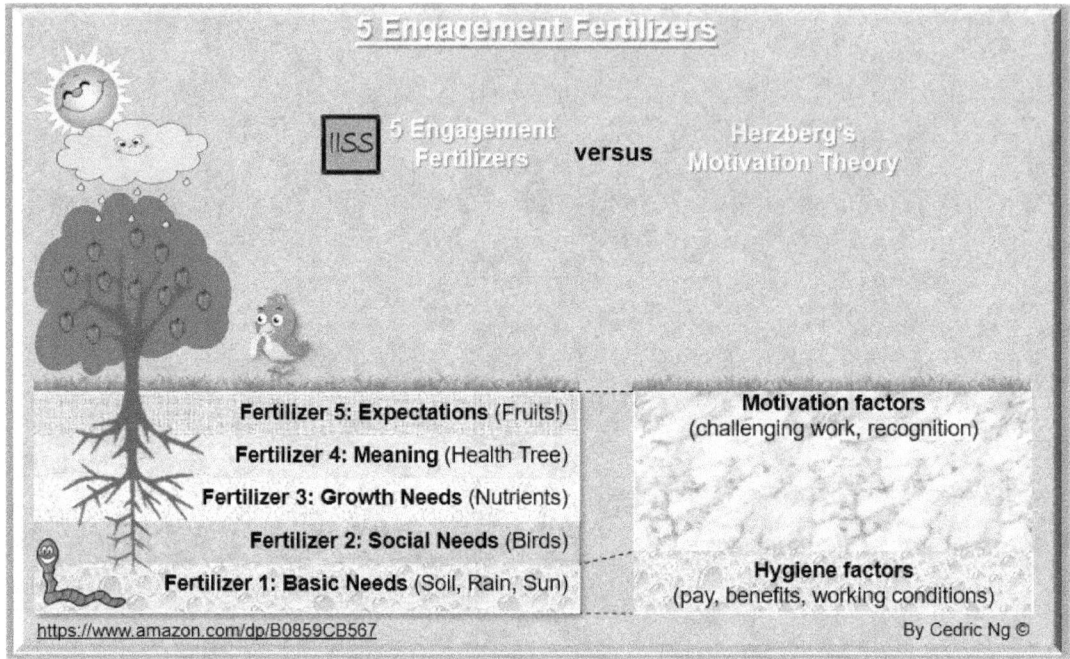

The "IISS model's Engagement Fertilizers" is similar to "Herzberg's Motivation Theory":
- **"Herzberg's Hygiene factors"** is similar to Fertilizer 1 (Basic Needs)
- **"Herzberg's Motivation factors"** is similar to Fertilizer 2 (Social Needs), Fertilizer 3 (Growth Needs) and Fertilizer 4 (Meaning) and Fertilizer 5 (Expectations).

3.1.3 Self-Determination Theory

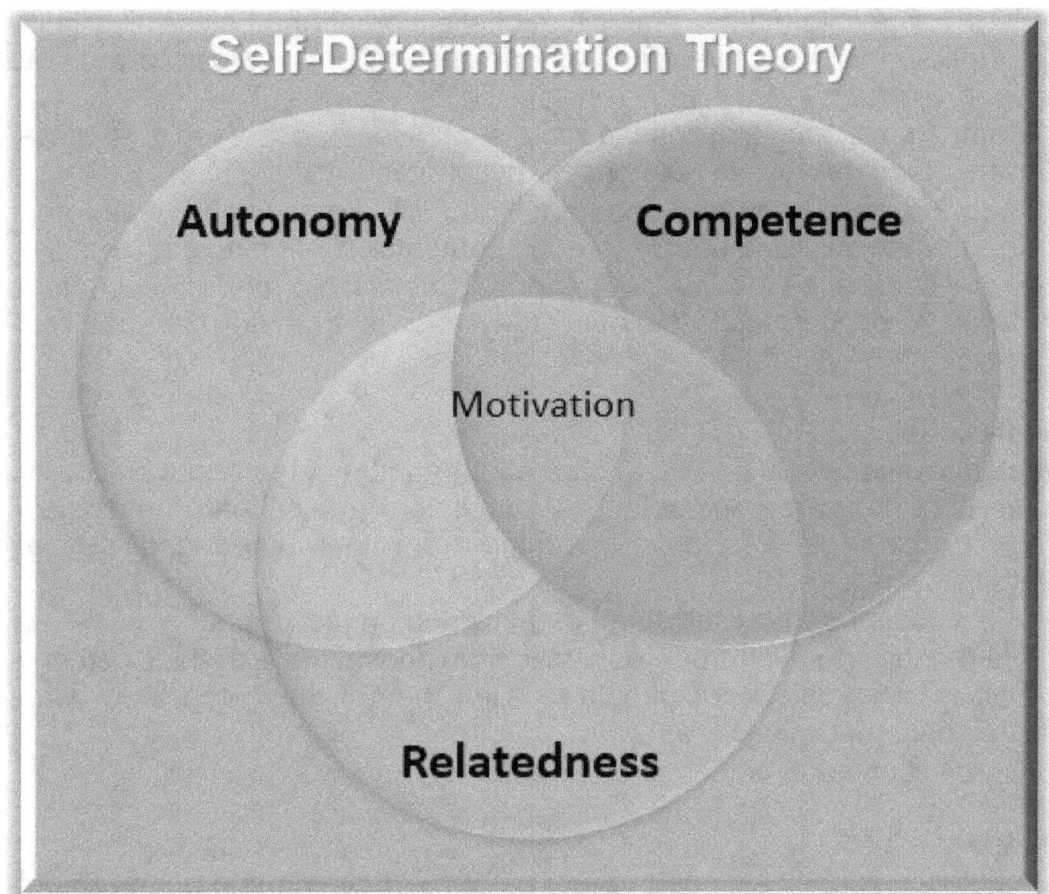

In the Self-determination theory, there are three universal psychological needs (autonomy, relatedness, competence). Unlike Maslow's needs, these three basic needs are not hierarchical or sequential. When the three basic psychological needs are satisfied in the workplace, people will be motivated. [1]

Autonomy

Autonomy is people's need to perceive that they have choices. The way leaders frame information and situations either promotes the likelihood that a person will perceive autonomy or undermines it. Ways to promote autonomy include: [1]

- frame goals and timelines as ways to assure a person's success, rather than as ways to hold people accountable.
- motivate people with the opportunity to fulfill a meaningful goal, rather than by incentivizing people through competition (winning a prize).
- Don't pressurize people to perform. Sustainable performance is a result of people acting because they choose to – not because they have to.

Relatedness

Relatedness is people's need to care about and be cared about by others, to feel that they are contributing to something greater than themselves. Ways to help people derive meaning from their work and deepen relatedness include: [1]

- Ask people how they fell about an assigned project or goal.
- Help people to align the organizational work values with their personal values. It is difficult for people to align them if they don't know what their personal values are.
- Connect people's work to a noble purpose.

Competence

Competence is people's need to feel a sense of growth and achievement. Ways to ignite people's desire to grow and learn include: [1]

- Make resources available for learning.
- Set learning goals – not just the traditional results oriented and outcome goals. At the end of each day, instead of asking, "What did you achieve today?" ask "What did you learn today?

<u>References</u>
(1) Susan Fowler, HBR (2019) What Maslow got wrong about our psychological needs. HBR Guide to Motivating People. United States of America. Harvard Business School Publishing Corporation.

3.1.3.1 "Self-Determination Theory" vs "IISS Model's Engagement Fertilizers"

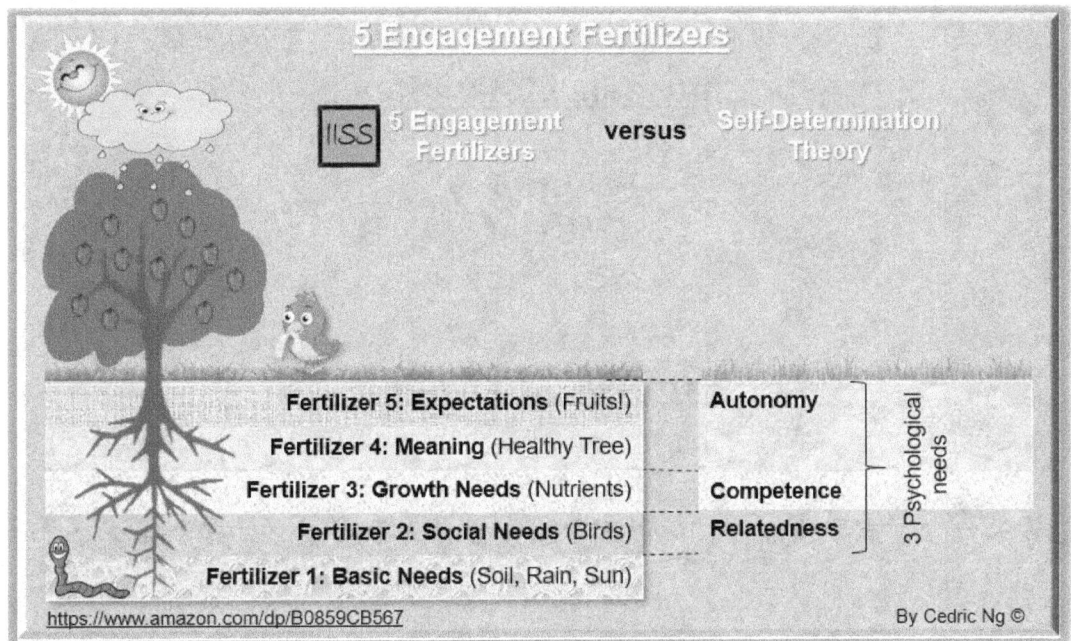

The IISS model's Engagement Fertilizers is similar to the Self-determination theory's three universal psychological needs (autonomy, relatedness, competence).:

- **Fertilizer 2 (Social Needs)** is similar to Self-determination theory's "Relatedness".
- **Fertilizer 3 (Growth Needs)** is similar to Self-determination theory's "Competence".
- **Fertilizer 4 (Meaning)** and **Fertilizer 5 (Expectations)** is similar to Self-determination theory's "Autonomy".

3.1.4 Autonomy, Mastery, Purpose Framework

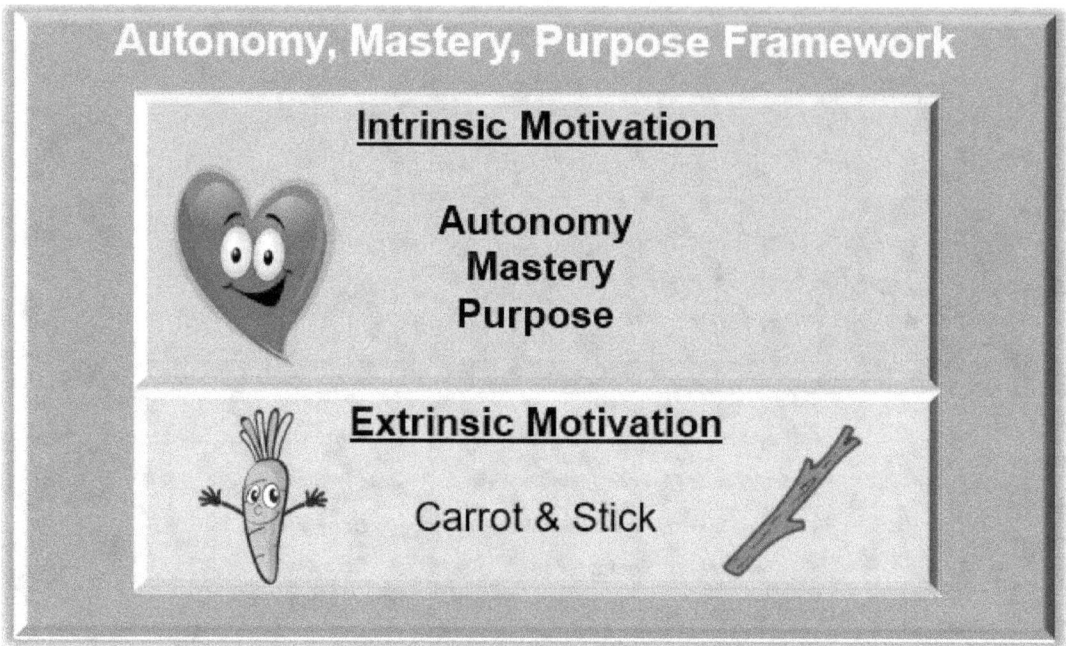

Money isn't the most effective motivator. Psychologist Edward Deci ran an experiment showing how incentivizing students with money to solve puzzles actually made them less interested in working on them after being paid. In contrast, another group of students who wasn't offered money, worked on the puzzles longer and with more interest. Deci's work uncovered the powerful difference between extrinsic motivation (motivation that comes from outside sources), and intrinsic motivation (motivation that comes from within yourself). [1]

Daniel Pink borrowed findings from research articles and contemporary management books to present his motivation framework. In his book, Drive, Pink illustrate why the traditional extrinsic motivation paradigm of carrots-and-sticks (reward and punishment) only works for simple straightforward tasks. Once people perceive that they are paid fairly, they need to be spurred by intrinsic motivations (Autonomy, Mastery, Purpose).

(i) Extrinsic Motivations

Extrinsic motivations are motivations that come from forces outside of the person experiencing them. Extrinsic motivators are things like bonuses and the threat of being fired (Carrot and stick). You have to pay employees enough, such that they are not focused on meeting basic needs and feel that they are being paid fairly. But once people's basic needs are met, extrinsic motivations become less effective. Once an employee gets a promised bonus for achieving their goals, where's the motivation to keep working hard? To further spur motivation, the employee needs to enjoy the activity and feel that it's important - they will need to be spurred by intrinsic motivations.

(ii) Intrinsic Motivations

Intrinsic motivations are motivations that originate from within a person rather than through external forces. For example, people who serve food to homeless folks are motivated by intrinsic desires. Lasting motivation comes from autonomy, mastery and purpose. [2]

Autonomy

Autonomy is the need to direct your own life and work. If managers want employees to be more engaged in what they are doing (as tasks become more complicated), then allowing employees autonomy (self-direction is better). Autonomy means you get to control what you do, when you do it, and who you do it with. According to Pink, autonomy motivates us to think creatively without needing to conform to strict workplace rules. Pink summarizes autonomy into four main aspects: time, technique, team and task. Many technology companies (e.g. Google) has benefited from numerous innovative product ideas as a result of allowing developers to pursue their own projects during work time. Providing the technology and freedom to work from home is another example of allowing staff more autonomy. [3]

Mastery

Mastery is the desire to improve. If you are motivated by mastery, you'll constantly seek to improve your skills through learning and practice. Someone who seeks mastery needs to attain it for its own sake. For example, an athlete who is motivated by mastery might want to continuously improve his running timing. [3]

Purpose

Purpose is the desire to work towards something worthwhile. People who find purpose in their work unlock the highest level of motivation. Purpose is what gets you out of bed in the morning and into work without grumbling. People who have purpose are motivated to pursue the most difficult problems. Encouraging people to find purpose in their work by connecting their personal goals to organizational targets, can win not only their minds, but also their hearts. [3]

References
(1) Janet Choi (2017) *How to Boost Productivity: Autonomy, Mastery, and Purpose*
https://blog.deliveringhappiness.com/the-motivation-trifecta-autonomy-mastery-and-purpose (13 February 2020)
(2) toggl (2019) *Autonomy Mastery Purpose: How to Use Author Daniel Pink's Framework to Successfully Motivate Your Team* *https://toggl.com/blog/autonomy-mastery-purpose* (13 February 2020)
(3) Mindtools (2020) *Pink's Autonomy, Mastery and Purpose Framework*
https://www.mindtools.com/pages/article/autonomy-mastery-purpose.htm (13 February 2020)

3.1.4.1 "Autonomy, Mastery, Purpose Framework" vs "IISS Model's Engagement Fertilizers"

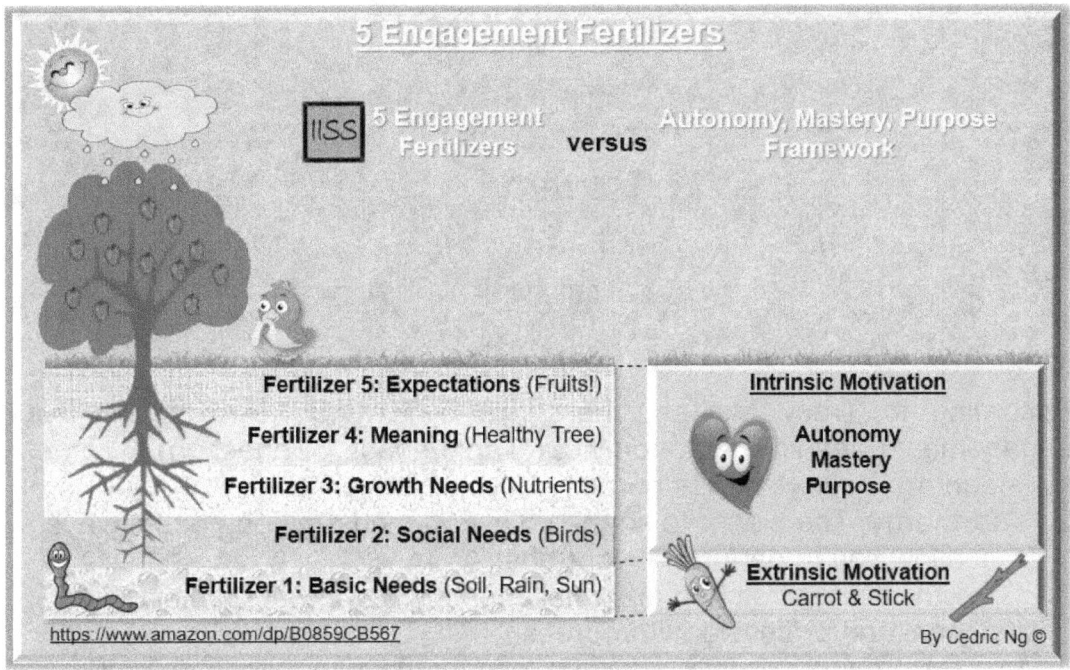

The IISS model's Engagement Fertilizers is similar to Daniel Pink's "Autonomy, Mastery, Purpose Framework":

- **"Daniel Pink's Extrinsic Motivation Factors"** is similar to Fertilizer 1 (Basic Needs)
- **"Daniel Pink's Intrinsic Motivation Factors"** is similar to Fertilizer 2 (Social Needs), Fertilizer 3 (Growth Needs), Fertilizer 4 (Meaning) and Fertilizer 5 (Expectations).

3.1.5 Engagement MAGIC model

Engagement MAGIC model
- **M**eaning
- **A**utonomy
- **G**rowth
- **I**mpact
- **C**onnection

According to Tracy Maylett, there are 5 MAGIC keys of employee engagement (Meaning, Autonomy, Growth, Impact, Connection). [1]
1. **Meaning**: Your work has purpose beyond the job itself.
2. **Autonomy**: The power to shape your work and environment in ways that allow you to perform at your best.
3. **Growth**: Being stretched and challenged in ways that result in personal and professional progress.
4. **Impact**: Seeing positive, effective, and worthwhile outcomes and results from your work.
5. **Connection**: The sense of belonging to something beyond yourself.

References
(1) Tracy Maylett (2018) The 5 Keys to Employee Engagement https://www.tlnt.com/the-5-keys-to-employee-engagement/ (13 February 2020)

3.1.5.1 "Engagement MAGIC model" vs "IISS Model's Engagement Fertilizers"

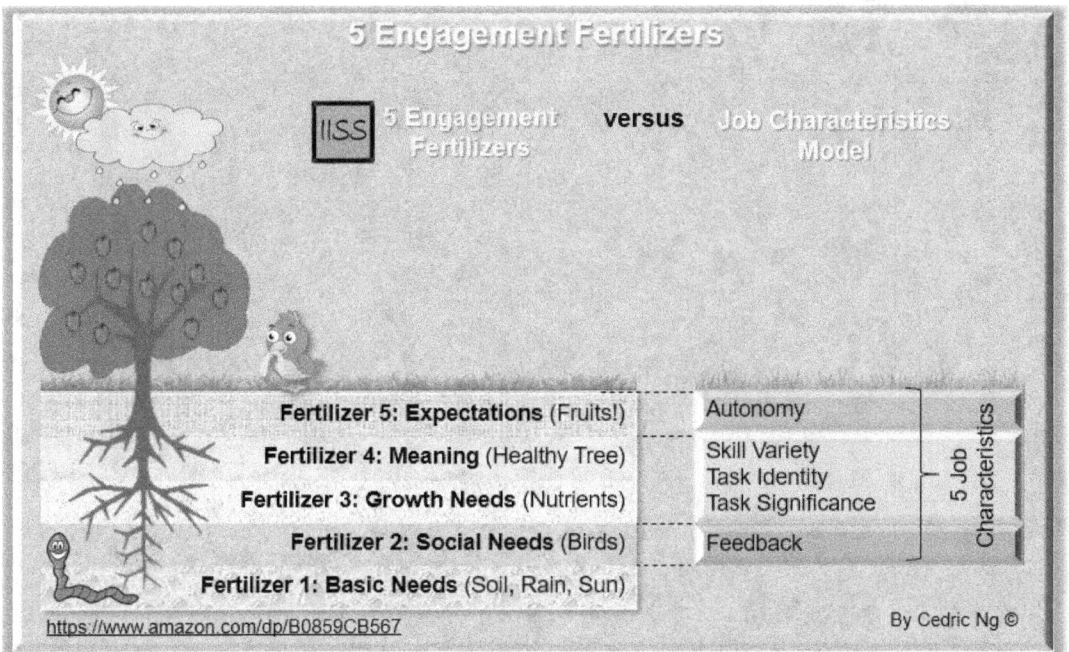

The IISS model's Engagement Fertilizers is similar to the Engagement MAGIC model's 5 keys for engagement (Meaning, Autonomy, Growth, Impact, Connection).:

- **Fertilizer 2 (Social Needs)** is similar to Engagement MAGIC model's "Connection".
- **Fertilizer 3 (Growth Needs)** is similar to Engagement MAGIC model's "Growth" and "Impact".
- **Fertilizer 4 (Meaning)** and **Fertilizer 5 (Expectations)** is similar to Engagement MAGIC model's "Meaning" and "Autonomy".

3.2 Engagement Fertilizer 1: Basic Needs

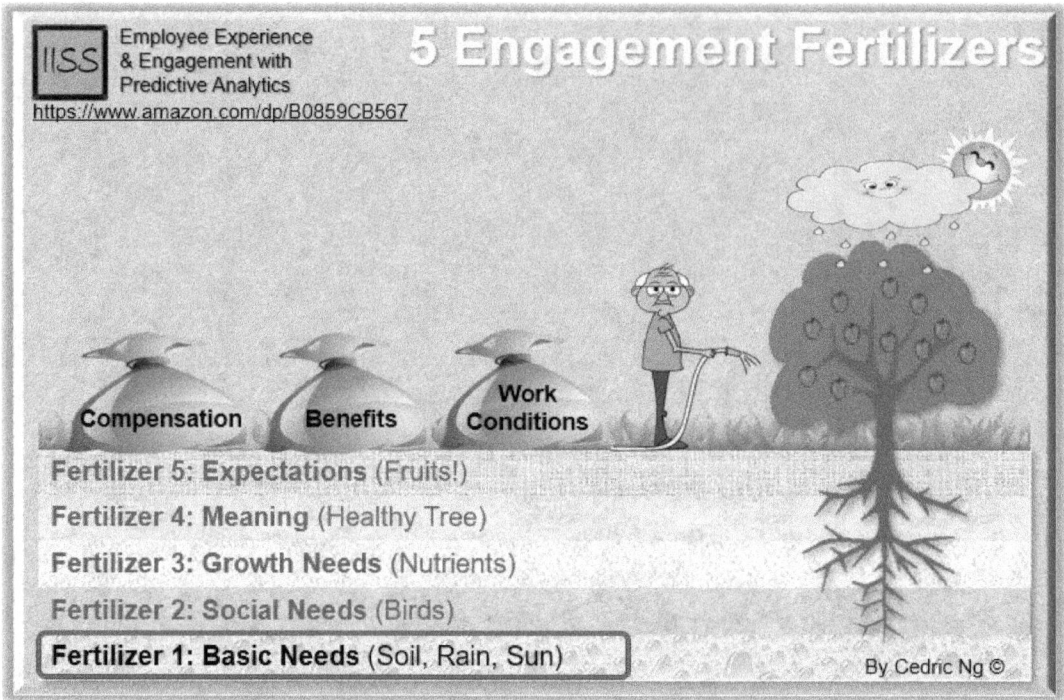

Engagement Fertilizers are needed to create the fertile soil for great employee experience & engagement. In order for your tree to survive and grow, it needs engagement fertilizer 1 (soil, rain, and sun).

Similarly, to have satisfied employees, you need to meet their basic needs for salary, benefits, and work conditions. People become dissatisfied when they feel that the basic needs they should be entitled to, is not there, or has been taken away. However, constantly introducing more and better basic needs doesn't increase job satisfaction or performance because of the "Principle of Adaptation". When people move to a higher income level or higher living standard, they adjust and become dissatisfied again. If you buy employee satisfaction by increasing basic needs and perks, the price always goes up. Money and perks matter - it's hard to be engaged when you feel underpaid or shortchanged. But money isn't why people love their jobs, and although perks are important, they don't engage.

3.2.1 Compensation

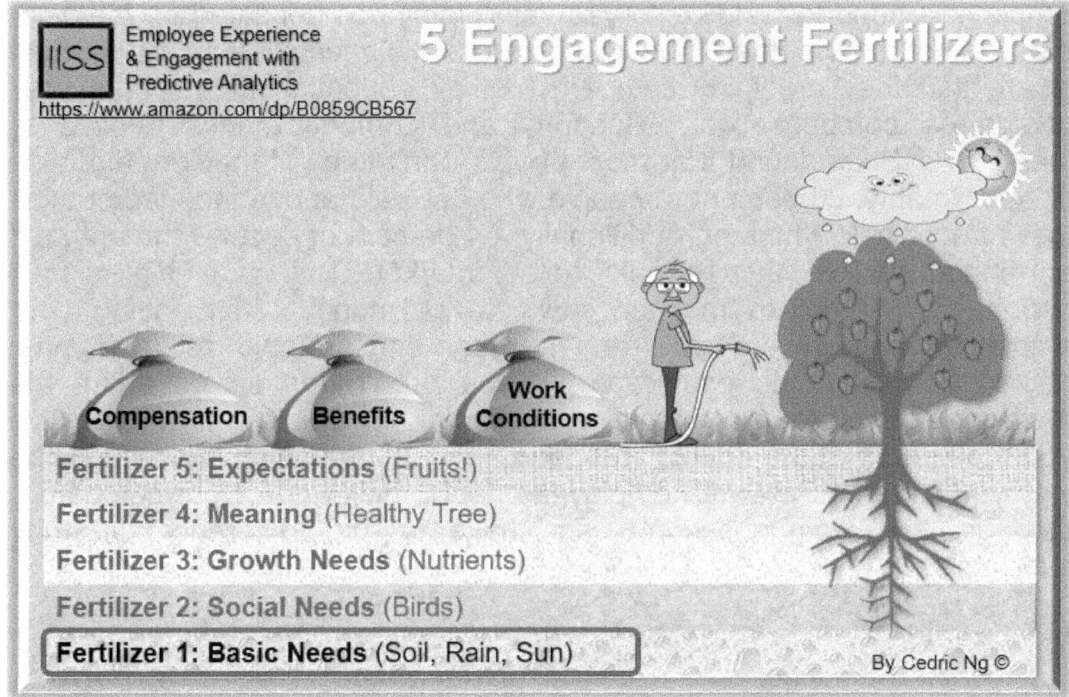

Money, perks and working conditions matter - it's difficult to be engaged when you feel short-changed.

3.2.1.1 Perception of pay

Employees need to feel like they're being fairly compensated for the work they do, otherwise they'll be disengaged. Help employees to become aware that they are fairly compensated by educating them about your company's compensation philosophy and competitiveness relative to market. PayScale found that most people don't know if they're paid fairly or not. Notably, 80 percent of people who are paid above the market think they're at or below market. Additionally, 64 percent of people who are paid at market believe they're paid below market. This is a big missed opportunity — not to mention a waste of money! Just paying your employees well isn't enough — the message they are paid well (and fairly) has to be clearly communicated. If you aren't transparent about the methodology you're using to make pay decisions, employees will not give you the benefit of the doubt. The good news is that small changes in the way you communicate with employees about their value, your company's strategy and compensation, can make a big difference in their satisfaction.[1]

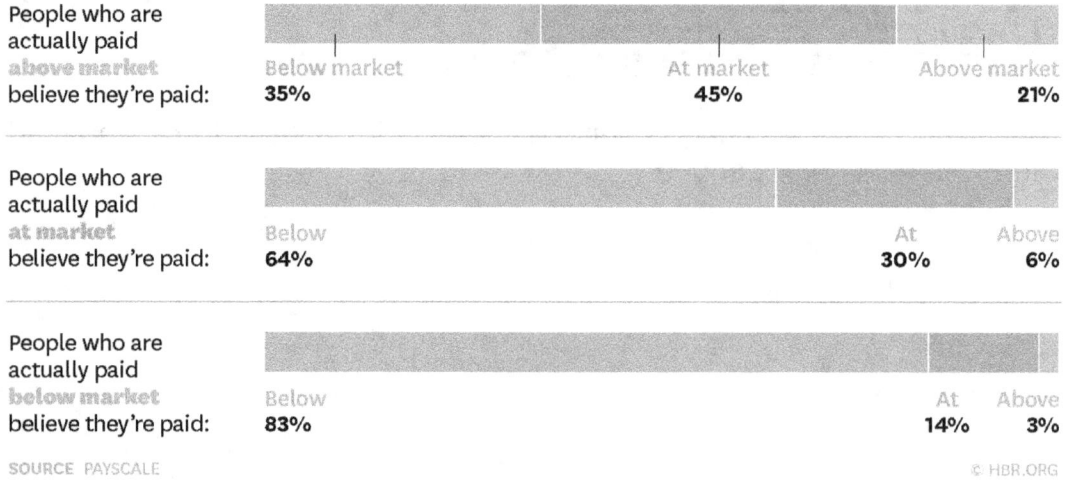

SOURCE PAYSCALE © HBR.ORG

Reference
(1) Payscale (2018) Employee turnover: 5 ways to spot flight-risk employees and what to do to retain them. https://www.payscale.com/compensation-today/2018/01/employee-turnover-5-ways-spot-flight-risk-employees-retain (26 December 2019)

3.2.1.2 Compensation for Salespeople

3.2.1.2.1 One rewards strategy does not fit all

A reward system that increased performance in one team, can led to disgruntled employees in another team. By using the same reward systems, you're managing your products and people as if they are the same, when they are not. Sales compensation plans that considers product and individual differences usually perform better. This section covers how organizations can improve sales.

3.2.1.2.2 Motivating core-performers

Core performers usually form the largest part of the sales force, and organizations cannot achieve their targets if they're not motivated. Strategies motivate core-performers include. [1]

Prizes

The problem with contests is that high-performers usually win them. Thus, core performers are not motivated to increase their efforts. The key to design effective sales contests that motivates both high-performers and core-performers is to offer gifts (not cash) for the lower-level prizes that can be seen as equal, or even superior to the top prizes. Suppose, in a Sales contest, the first prize is an overseas vacation to Canada, while the second prize is an overseas vacation to Mexico. If the Mexico trip cost lesser than the Canada trip, core performers can rationalize their prize by saying, "as I've went to Canada already, the overseas vacation to Mexico is more valuable to me". The lower-level prize must have some quality that the higher-level one does not. Core performers work harder and perform better in contests gifts prizes, than contests with cash prizes. Furthermore, their increased effort does not come at the expense of lesser effort from high-performers or under-performers. [1]

Multi-tier targets

A research by Thomas Steenburgh and Michael Ahearne found that core performers given 3-tier targets, outsold core performers given 2-tier targets. In contrast, multi-tier targets did not result in significant difference in performance for high-performers and under-performers. There is no significant difference in performance for high-performers because the top tier is attainable for them regardless of the number of targets. Under-performers are unaffected by the extra stepping stone, as they usually just aim for and are satisfied with achieving the first-tier target. [1]

3.2.1.2.3 Motivating under-performers

Under-performers may include new hires who needs training and complacent senior salespeople, and people who are less talented and motivated. Most under-performers performance can improve if the right incentives are in place. [1]

Pace-setting bonuses

A study by Thomas Steenburgh found that removing quarterly bonuses from incentives, and keeping only an annual bonus, decreases the performance (as measured by the revenues they generate) of under-performers, core-performers and high-performers by 10%, 4% and 2% respectively. Similarly, weaker students need periodic quizzes throughout the semester to keep them on track. In the absence of such mechanisms, they perform poorly on comprehensive exams. By contrast, strong students – like high-performers salespeople – make an effort independently and have less need for intermittent goals. [1]

Natural social pressure

Having a pipeline of new sales talent puts social pressure on low-performing salespeople - this is referred to as the "man on the bench" effect, because it is similar to the pressure that second-string quarterbacks place on starters in football. Thomas Steenburgh and Michael Ahearne found that salespeople in districts with a bench player perform 5% better than those without one. The greatest increase in performance takes place in the under-performers group. In the long run the overall increase in revenue easily outweighs the additional costs associated with hiring bench players. [1]

Program-induced social pressure

Programs that put social pressure on under-performers may work in competitive and transparent organizational cultures. But it should be sensitive and implemented with care. Some companies induce social pressure by posting sales numbers in ascending order from under-performers to high-performers. Other companies publicly list each of its salespeople in one of the three categories: starters, benchwarmers, and the penalty box. [1]

3.2.1.2.4 Motivating high-performers

No ceiling on commissions

Colin Camerer studied New York City cabdrivers and found that most cabdrivers quit for the day once they reached their target ("income targeting"). A study by Sanjog Misra and Harikesh Nair on a large US contact lens manufacturer, found that removing sales commission caps increased revenue by 9%. By placing caps on commissions when salespeople are hot, companies encourage high-performers to quit selling – just as cabbies go home early on rainy days, when their hourly earnings are highest. Companies may be better off removing sales commission caps, so that high-performers are incentivized to work more intensively during high demand periods. [1]

Overachievement commissions

Overachievement commissions are higher commissions rates that kick in after quotas are met. Thomas Steenburgh's found that removing overachievement commissions from a compensation plan reduces high-performers' sales by 17%.[1]

Multiple winners

Michael Ahearne found that contests with multiple winners boost sales effort and performance better than contests with winner-takes-all prize structures. [1]

3.2.1.2.5 Designing incentives for greater leverage

As salespeople at different points on the performance curve will respond to different incentives, the first step for any company is to get a clear understanding of its own performance curve. To derive an organization's performance curve, calculate each person's performance against sales targets and then create a histogram of those data - you'll be able to see whether your company's curve is normal, under-performers heavy, or high-performers-heavy. The shape of a company's performance curve will suggest which incentives will give you the most leverage. If you have a disproportionate number of under-performers, you'll want to focus first on pace-setting bonuses and natural social pressure, for example. [1]

Reference
(1) Thomas Steenburgh and Michael Ahearne, HBR (2019) *Motivating Salespeople: What really works. HBR Guide to Motivating People.* United States of America. Harvard Business School Publishing Corporation.

3.2.2 Benefits

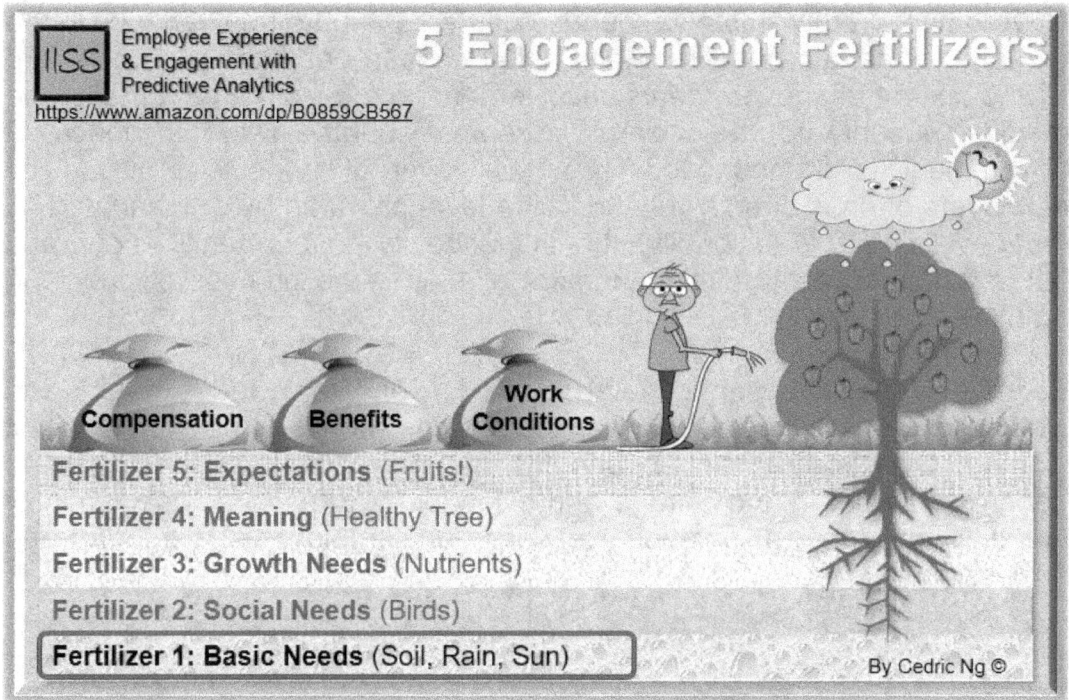

Employee benefits represent a considerable portion of the total rewards package and thus a significant expense to the organisation. The importance of the benefits function is highlighted in its role of attracting and retaining employees and fulfilling employee needs. Regardless of the country, benefits serve the same essential functions. Benefits supplement remuneration by providing employees with a level of security related specifically to health and welfare, retirement and time off.

While employee needs may be similar around the world, the degree to which those needs are met through other courses varies by country. Governmental programs and country culture contribute to a variance in the demand for certain benefits. Many differences in benefits practices relate to the extent of governments involvement in providing retirement and medical security to its citizens. In some countries, the influence of government is pronounced. Thus, the level of benefits provided by employers often is directly related to the level provided by or mandated by the government. Additionally, the influence of government, labour and other factors tends to either augment or hinder the process of providing employee benefits.

Benefits are a core part of total rewards. Benefits includes health and welfare plans, retirement plans and programs providing pay for time not worked. Overtime, employee benefits have evolved from basic "fringe benefits" of insurance coverage and a few perquisites to a wide range of benefits to strike a balance between an employee's personal and professional life.

3.2.2.1 Why give Benefits?

Relationship between "Herzberg two-factor theory" & "Employee benefits positioning"

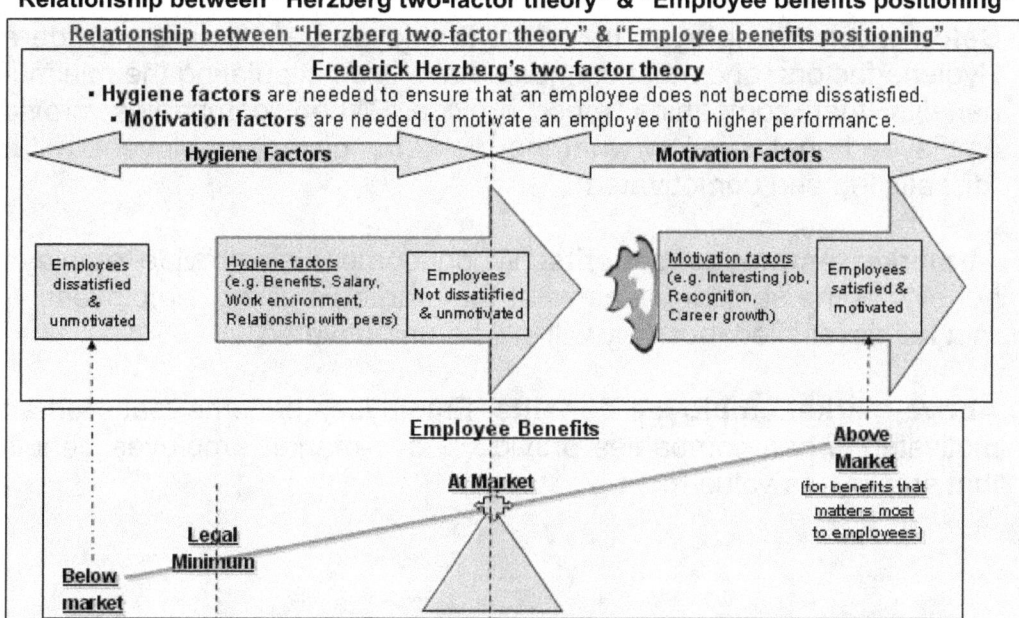

Source: https://www.amazon.com/Reviewing-Employee-Benefits-using-model-ebook/dp/B07FXSYKB2/

Organizations that offers better than average benefits may pay less salary and still have motivated, contributing employees. Frederick Herzberg's motivation-hygiene theory states that there are certain factors in the workplace that cause job satisfaction, while a separate set of factors cause dissatisfaction.

- **Hygiene factors**: Hygiene factors are needed to ensure that an employee does not become dissatisfied. Examples of Hygiene factors are employee benefits, salary, work environment.

- **Motivation factors:** Motivation factors are needed to motivate an employee into higher performance. Examples of Motivation factors are interesting job, recognition, and career growth.

The Herzberg's two-factor theory can be linked to employee benefits market positioning.

- **Below-market employee benefits**: Employee benefits are considered Hygiene factors, and most countries have laws stipulating the minimum benefits that companies must provide. When companies provide employee benefits below what the market is giving, employees will be "dissatisfied and unmotivated".

- **At-market employee benefits**: When companies provide employee benefits at the same level as what the market is giving, employees will "not be dissatisfied, but they will still be unmotivated".

- **Above-market employee benefits**: Employees become "satisfied and motivated" when companies provide above-market employee benefits that employees value most.

3.2.2.2 Benefits to drives engagement

Benefits for gig contractors

The gig economy & freelancing has become the new talk of the town. Many talented people have decided to leave the traditional workforce for freelance work in their area of expertise. According to Intuit by 2020 40% of American workforce will be independent contractors to live a more flexible work life. This emerging trend is becoming a challenge for HR Managers. HR have to start identifying roles for transitioning to the gig economy, integrate gig contractors by offering them a sense of belonging (e.g. social security, basic medical benefits), and establish a pool of go-to freelancers that they can rely again and again. [1]

In Amazon, work-from-home Customer Service Agents gets life and disability insurance, dental and vision insurance, medical insurance, and education benefit which pre-pays 95 percent of courses. [2]

Flexible work options

Employees need a life outside of work. Allow work-from-home, provide flexible hours options, and ban after-work emails and messages, so employees have the freedom to enjoy life to the fullest, but still get their work done. In 2016 South Korea submitted bill to ban bosses from bothering their staff at home. The bill bans firms from sending employees work related messages by telephone, text, social media or via mobile messaging apps after official working hours. The document specially references KakaoTalkm a chat app used by 80 percent of South Koreans. [3]

Amazon announced on 6 April 2017 that it is looking for 5,000 Customer Service Agents who can work from home. Those who work more than 20 hours a week get life and disability insurance, dental and vision insurance, medical insurance, and education benefit which pre-pays 95 percent of courses. Working from home is crucial to work-life balance. By not commuting or having to wake up early, you're saving not just money and time but also energy, which is important to meet work and family obligations. [4]

Low cost benefits

You don't need to break the bank to offer expensive employee benefits. A study by Kerry Jones, Marketing Manager at Fractl, found that after health insurance, employees place the highest values on benefits that are relatively low cost to employers, such as flexible hours, more paid vacation time, and work-from-home options. [5]

Snacks

Free coffee is nice, but a pantry full of snacks is even better. About 67 percent of employees who receive free food at work are very happy with their jobs. **The key to happiness at work is free snacks -** According to a new survey by grocery-delivery service Peapod, 56% of full-time employees are "very" happy with their current job, that number jumps to 67% among those who have access to free food. The same study found that this perk is more important to millennials than to other Groups. 48% of respondents (especially millennials) said that if they were looking for a new job, they would weigh company perks, including availability of snacks, in their decision. [6]

Wellness

Workplace wellness programs can increase productivity, reduce stress, increase engagement, and boost your bottom line. When employees see that their employer values them, it will make them more inclined to put back into the company. [4] Examples of wellness activities are lunchtime yoga sessions, subsidized healthy snack delivery, Refrigerators and cupboards stocked with fruits for the entire office, corporate gym memberships, Hydraulic Tables for office employees, Dependent benefits, Free family counselling hotline, Health/Stress Talks, Under-Desk cycling exercise machine, etc. Quantum Workplace found that employees who work at organizations that provide healthy vending options are 10 percent more likely to be engaged. [7]

References
(1) zyksham (2019) *The Gig Economy, and why it Matters* https://zyksham.blogspot.com/2019/03/gig-economy-and-why-it-matters.html?showComment=1556336036464&m=1&fbclid=IwAR3QMe5wgHLoTy0A8hsPBWwUTVK7g4nU-r02OQB8VsNlkPdo5ghho5cZuUQ *(3 October 2019)*

(2) Maricar Santos and Joseph Barberio (2019) *Amazon Is Hiring 3,000 Work-from-Home Employees with Full Benefits* https://www.workingmother.com/amazon-is-looking-for-5000-work-from-home-employees?src=SOC&dom=fb *(3 October 2019)*

(3) Rapper.com (2016) *South Korea mulls law to keep office out of the home* https://www.rappler.com/world/regions/asia-pacific/137499-south-korea-mulls-law-keep-office-out-home?utm_content=buffer18387&utm_medium=social&utm_source=facebook.com&utm_campaign=buffer&fbclid=IwAR1DfcafFiUgxPWSHbZ7tEjAxrgfdqKUq-jJlM3cFLypcHjjZ-oTCh63SMI *(3 October 2019)*

(4) Source: Maricar Santos and Joseph Barberio (2019) *Amazon Is Hiring 3,000 Work-from-Home Employees with Full Benefits* https://www.workingmother.com/amazon-is-looking-for-5000-work-from-home-employees?src=SOC&dom=fb&fbclid=IwAR24UwwWh3OiXGEitouySPGelmm-rmhtkKdveYPdzD2Fpe9lrdKvEomY2FI *(3 October 2019)*

(5) HBR (2019) *It takes more than just extrinsic rewards to inspire people.* HBR Guide to Motivating People. United States of America. Harvard Business School Publishing Corporation.

(6) Hadley Malcolm (2015) *The key to happiness at work is free snacks* https://www.usatoday.com/story/money/2015/09/16/study-says-snacks-affect-happiness-at-work/72259746/ *(3 October 2019)*

(7) Emil Shour (2019) *59 Awesome Employee Engagement Ideas & Activities for 2019* https://www.snacknation.com/blog/employee-engagement-ideas/ *(3 October 2019)*

3.2.3 Work Conditions

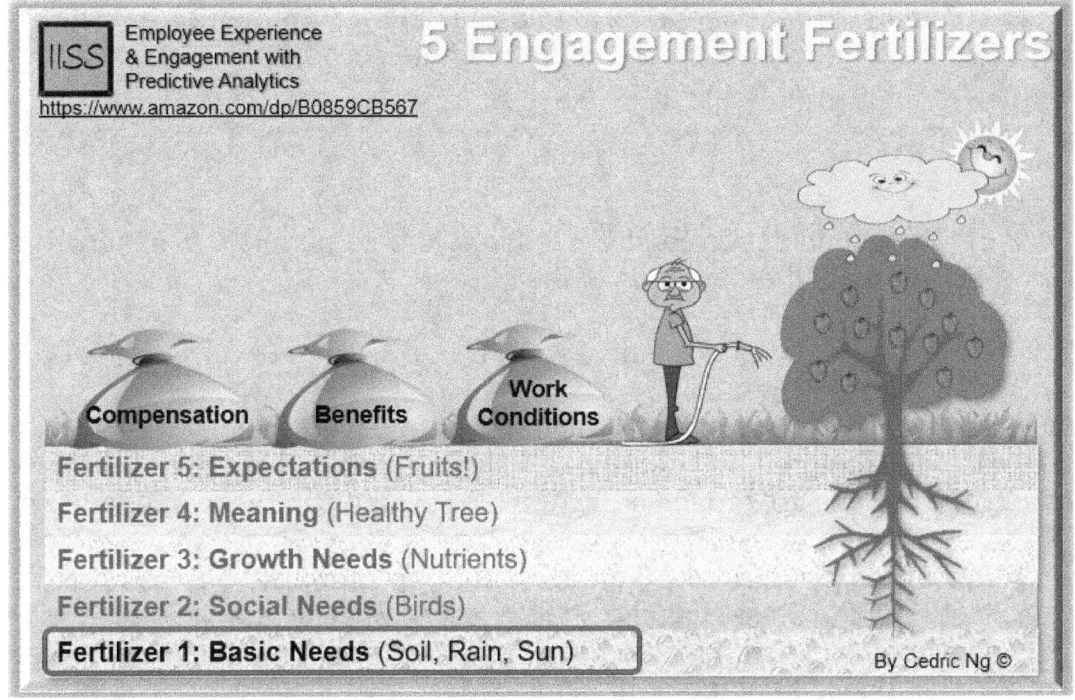

Work conditions refers to: commuting time, job security, physical working environment, working hours, work resources & processes.

Commuting time

Long commute is detrimental to job satisfaction for the same reasons as long hours are, and in addition, long commutes tend to be unpleasant in themselves and hard to use productively.

Job security

Lack of job security leads to stress as it is affects people's confidence in being able to meet the demands of their lives.

Physical working environment

Office spaces plays a role in attracting and retaining talent. Redesigning the office environment and can make it more enjoyable to be in. It doesn't have to be expensive, repainting and adding plants can make a difference.

- **Align office design with values.** When an organization says that its values is fun, these values should be physically manifested in the workplace. E.g. workplace attire can be informal, meeting rooms colorful, fun wall photographs, employees have flexible working hours, etc. [1]

- **Quiet spaces for head-down work.** Workplace design can be a powerful force for enhancing or diminishing employee engagement. It is difficult for employees to concentrate in a sea of faces when the cubicles are removed and everyone is sitting at long tables. Despite popular belief, open spaces are not spurring useful communication. Some companies try to get around the pitfalls of open offices by providing a range of different workspaces. While a range of options is better than asking people to work in chaos, it is insufficient. Research shows that people lose concentration every time they are interrupted, including when they have to pack up and move to a quieter place in the office. [2]

Working hours

Very long working hours crowd affects your non-work life, which makes you less satisfied, and make you perceive your job as conflicting with your other goals and needs.

Work resources & processes

The resources & processes should support productivity. Streamline processes where possible, ensure employees have the office stationery to work, ensure there is availability of meeting rooms with projector and telephone, etc.

<u>References</u>
(1) Source: Hadley Malcolm (2015) *The key to happiness at work is free snacks* <u>https://www.usatoday.com/story/money/2015/09/16/study-says-snacks-affect-happiness-at-work/72259746/</u> *(3 October 2019)*
(2) Sally Augustin, HBR (2019) Rules for designing an inspiring workplace. HBR Guide to Motivating People. United States of America. Harvard Business School Publishing Corporation.

3.3 Engagement Fertilizer 2: Social Needs

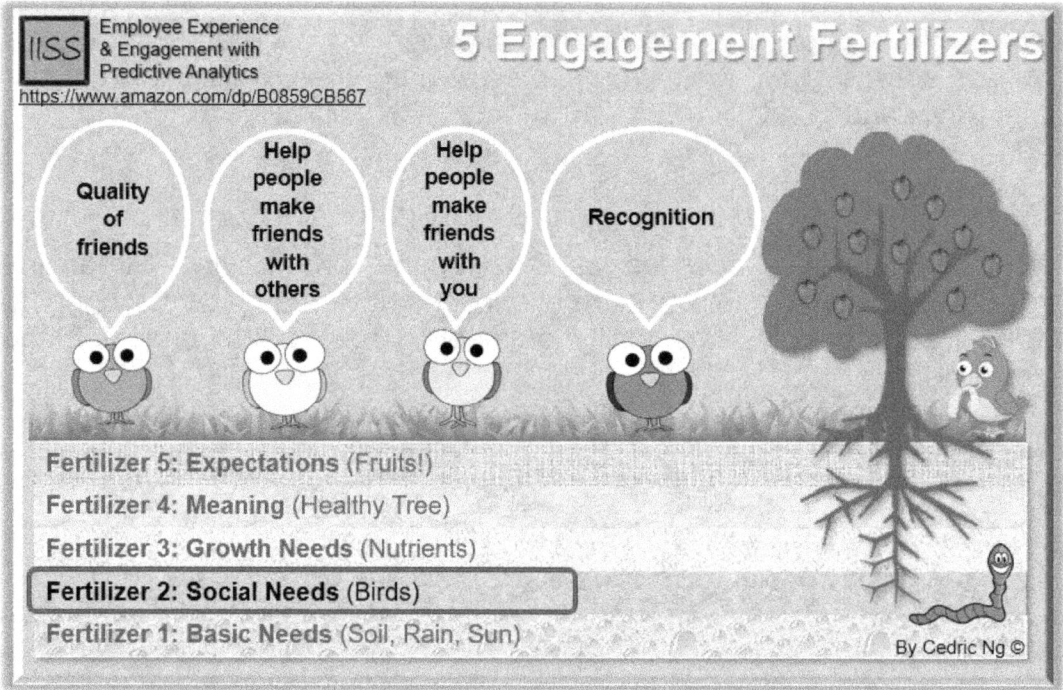

Engagement Fertilizers are needed to create the fertile soil for great employee experience & engagement. Birds helps to keep your fruit tree healthy by getting rid of the worms that eat the apples. Similarly, people need friends to keep them engaged. Mixing with disgruntled employees can destroy engagement. This is where engagement fertilizer 2 (Social Needs) comes in.

In 2013, the UN released some interesting findings. According to the report, of Earth's seven billion inhabitants, six billion have cell phones. Not surprising? Perhaps, until you learn that only 4.5 billion have proper sanitation. Is that right? One and a half billion people have cell phones but no toilet or sanitary waste disposal? What does this say about the importance of being connected? [1]

People have different social needs

What engages you may not engage another person. Some people love big birthday parties, while others hate it. An organization's role is to lay the conditions under which employees can choose how they will engage, then step back and let their social needs grow organically.

Organizations can till the soil and lay down the fertilizers for social needs through:
- Quality of friends
- Helping people make friends with others
- Helping people make friends with you
- Recognition

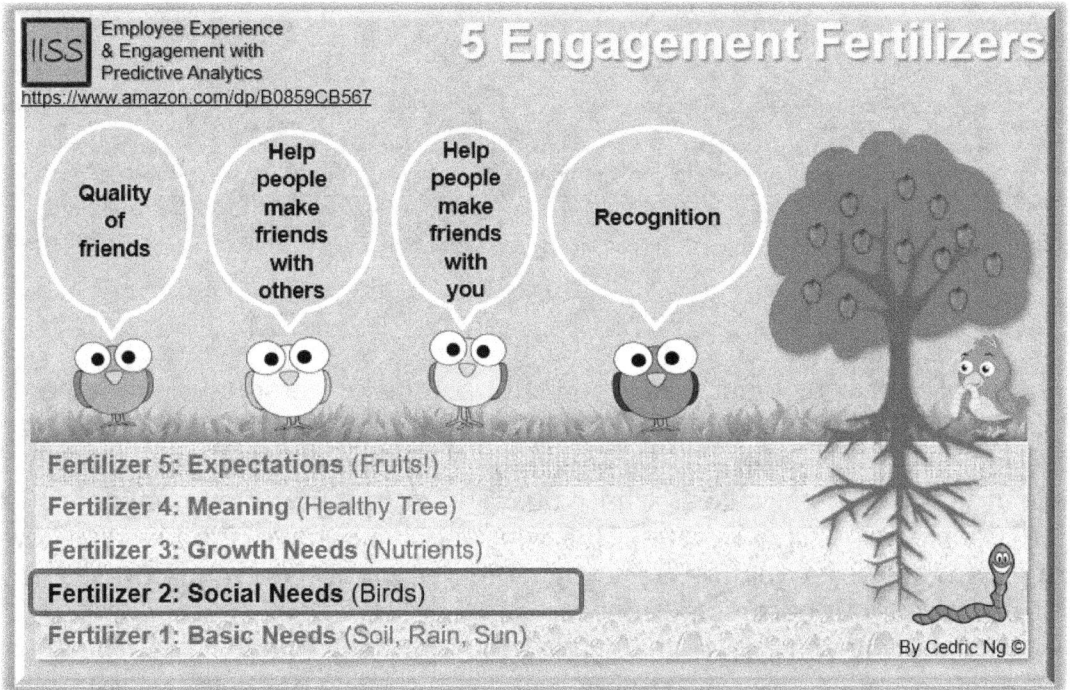

Reference
(1) Yue Wang (2013) More People Have Cell Phones Than Toilets, U.N. Study Shows https://newsfeed.time.com/2013/03/25/more-people-have-cell-phones-than-toilets-u-n-study-shows/
(3 October 2019)

3.3.1 Quality of friends

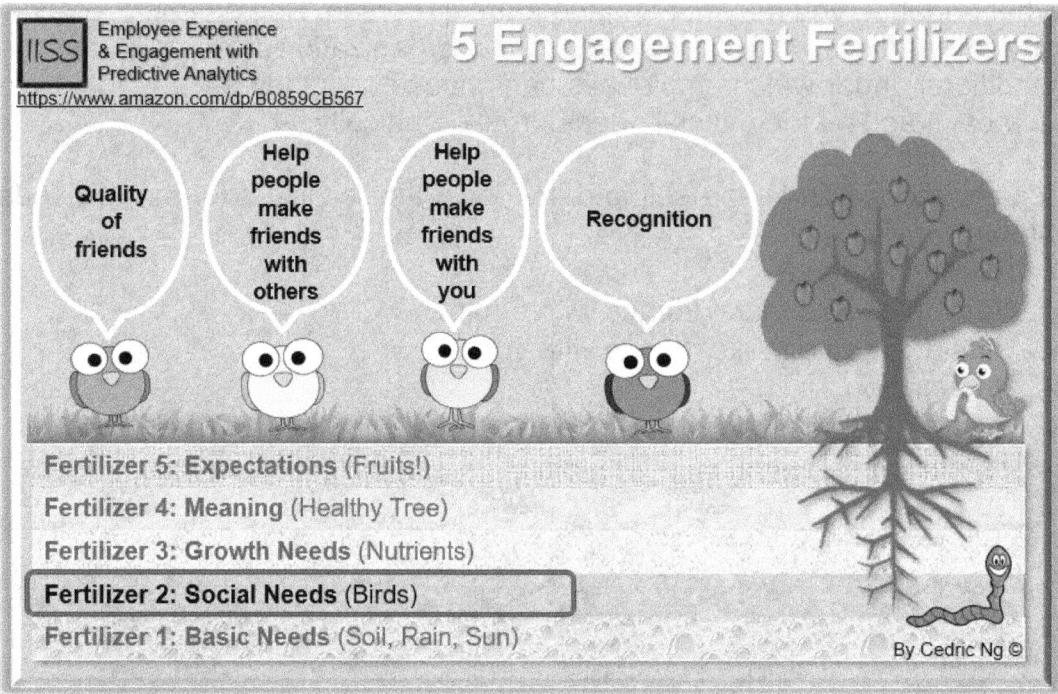

Review your quality of friends

The quality of your friends is a powerful source of personal engagement, but you have to do it right. Highly engaged people seldom keep low quality friends. Take a moment to assess your friends and write your thoughts about them. In most cases, what you'll find is that you stay connected to the people with high returns. But you may also have friends with low returns and you should be asking yourself why you keep the connection. List some of your current friends
- Explain why you chose to stay plugged into each one
- Explain how might you change your connecting behavior to get more quality friends

Power of multiple connections

One of the reasons it's so difficult to make a change is because of what the financial services world call "stickiness". Here's how stickiness works.: (1)

- When a customer opens a new checking account at a bank, it has gained a new customer. But the connection is not necessarily a strong one. If the customer is receiving only one product or service, such as a checking account, the chances of retaining the customer over the next 5 to 6 years is less than 40%.
- But if the customer is persuaded to add just one more service, such as credit card account or a certificate of deposit, customer retention increases to 70% because the bank has become stickier to the customer.
- If the customer adds a third product, retention jumps to over 90%. With that kind of stickiness, it's too hard to leave because there's a good return on connection (ROC). The customer has settled into a more intimate relationship with the bank. The same principle holds true for individuals in organizations. As you connect to the organization and its people in multiple ways, your engagement increases.

Reference
(1) Timothy R. Clark (2012) The Employee Engagement Mindset (11 January 2020)

3.3.2 Help people make friends with others

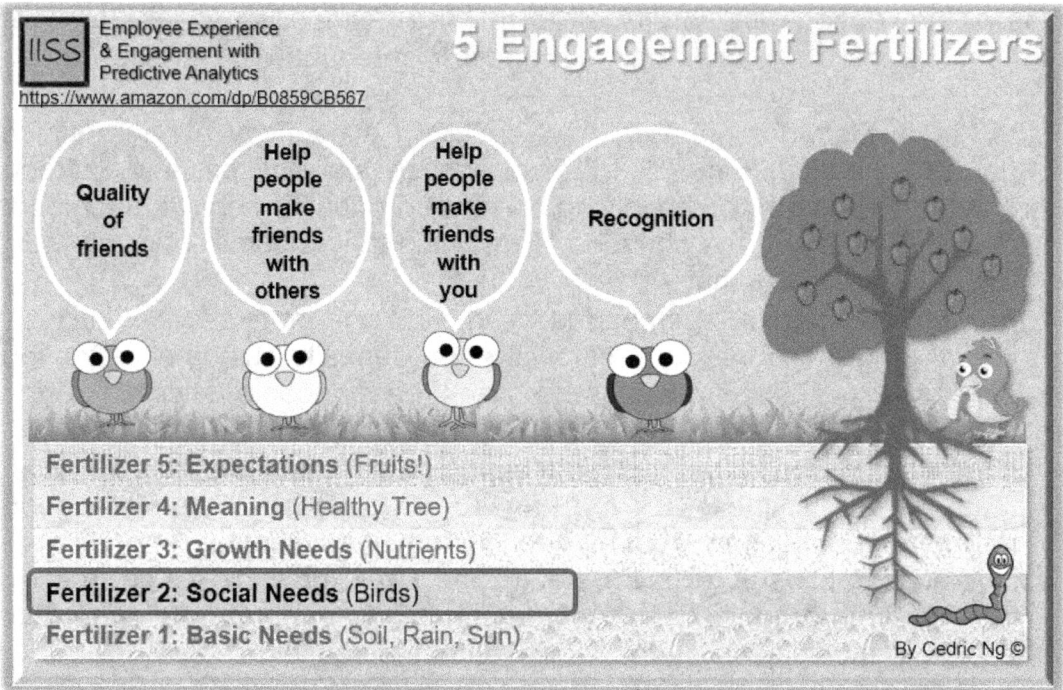

A survey found that employees with best friends at work are more engaged and resilient. Having a best friend at work affects social well-being, makes the job more fun, even when tasks get stressful. [1]

When your colleague is also your friend, it's easier to work. People don't feel emotion towards a business – they feel it towards other people. Thus, if you want to get your staff to invest emotionally into what they're doing, you need to help employees make friends at work.

Assign buddies

When bringing new employees into the organization, assign them to a buddy from your team who is not their supervisor. Have the buddy help the new employee build relationships with other employees. Set specific goals regarding whom the new employee will meet and when. This will accelerate the social connection rather than letting it evolve more slowly. (2)

Create downtime

Create some downtime for your employees to hang out--pockets of time where they'll be interacting with each other.

Culture fit

Cultural fit is more important that ability – skills can be taught but personality cannot. An employee who fits well with the team brings positive energy. An employee who is highly skilled but a poor team fit can be damaging. If you focus strictly on the capability of the employee, your team culture may flatter or worse become toxic, leading to poor morale and high turnover.

Permit social networking use

According to Robert Half Technology, over half of U.S. companies block social networks. Companies see social networks as productivity drainers and as a security issue. According to Cisco, more than half of young professionals wouldn't work at a company that blocked social networks. Social networks are a way of life for Gen Y employees and they want to be always connected to their friends and family, even during work hours. Since employees are answering e-mails and having calls outside of work hours, they should be able to be more social at work. If someone wants to be unproductive, they will find ways to do it, digital or otherwise. Your employees have smartphones at work, and are already on social media at work, so leverage it. [3]

Digital and mobile tools can improve productivity, and overall engagement. The proliferation of mobile devices and videoconferencing allows us to work anywhere and anytime. Limiting tools and access ends up negatively affecting morale and productivity. Online communication helps us to boost employee engagement by making it easy for employees to make friends at work. The era of being chained to the desk is over. People want freedom - that's what technology gives us. [4]

Avoid Social media addiction - however if we connect unwisely, it can lead to harmful addiction. Knowing yourself and applying discipline will help you fight digital addiction.

Organize Social events

Offsites, where employees are free to bond away from the stresses of office life, is one method to cultivate a workplace where more friendships are formed. Managers should engage their team through periodic happy hours, and company wide retreats. An employee who is surrounded by people who care and bond with them is more likely to feel engaged at work. Take a half day Friday to do something fun together (e.g. go on a scavenger hunt). These social events help people bond with others on the team who they don't interact with daily.

Reference
(1) Rever Team (2019) 15 Employee Engagement Activities For Your Team https://reverscore.com/employee-engagement-activities/ (2 October 2019)
(2) Timothy R. Clark (2012) *The Employee Engagement Mindset* (11 January 2020)
(3) Dan Schawbel (2012) *5 Ways to Retain Gen Y Workers* https://www.americanexpress.com/en-us/business/trends-and-insights/articles/5-ways-to-retain-gen-y-workers/?fbclid=IwAR0a2431isZHKS5JPeZGmU3GVXsp4pz5qX264XeD4CpdxoV5WfqL7t3Fvl4 (12 March 2020)
(4) *Lamont Exeter (2019) Three Ways to Digitize the Workplace to Boost Employee Engagement* https://www.ttec.com/articles/three-ways-digitize-workplace-boost-employee-engagement (3 October 2019)

3.3.3 Help people make friends with you

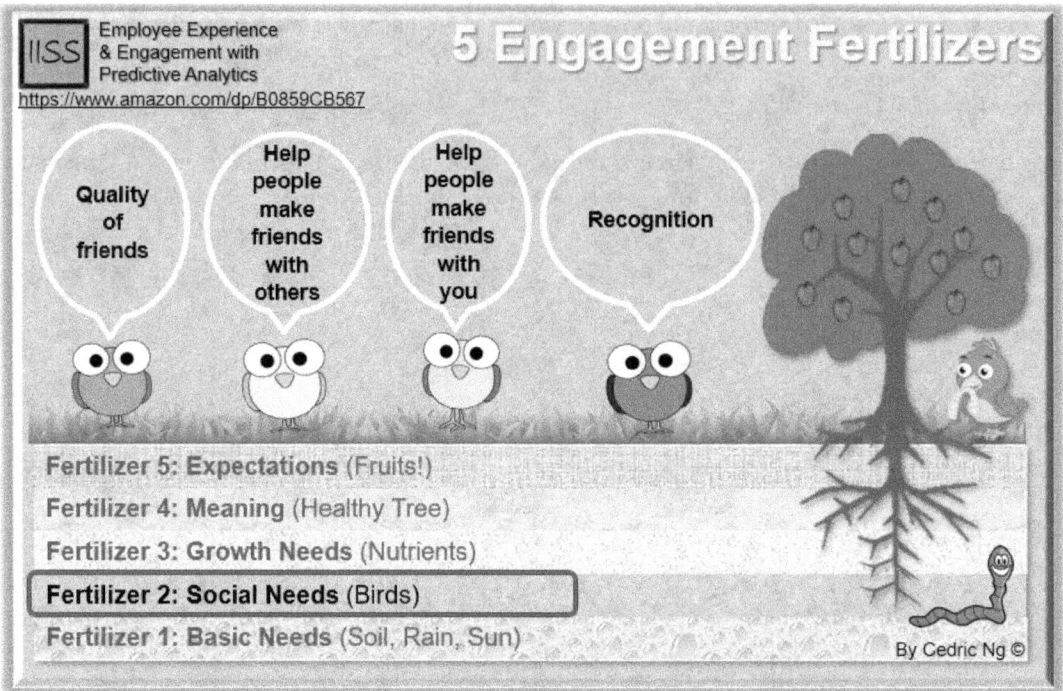

Other than helping people make friends with others, you also need to help people make friends with you - "you" as either their boss, or their colleague.

Stop by

Stop by and visit your team with no agenda in mind and no directive to deliver. You are just there to catch up on everyone's lives.

Have a one-one-one over coffee

You need to get up from your desk from time to time, anyway. Grab a team member for a quick walk and talk and you've just multitasked your way to a stronger connection. You'll not only get up to speed quickly with your department's goings on, you'll provide the valuable face-time your employees crave. Engagement increases as an individual gets more time with their bosses. Ask your employees whether they are happy at work and what you can do to make them happy. Don't wait for annual reviews to have this conversation. By communicating regularly with your employees, you will know what motivates them and the challenges that they need to overcome - this helps you reward their employees in ways that are meaningful to them, which can change over time.

3.3.4 Recognition

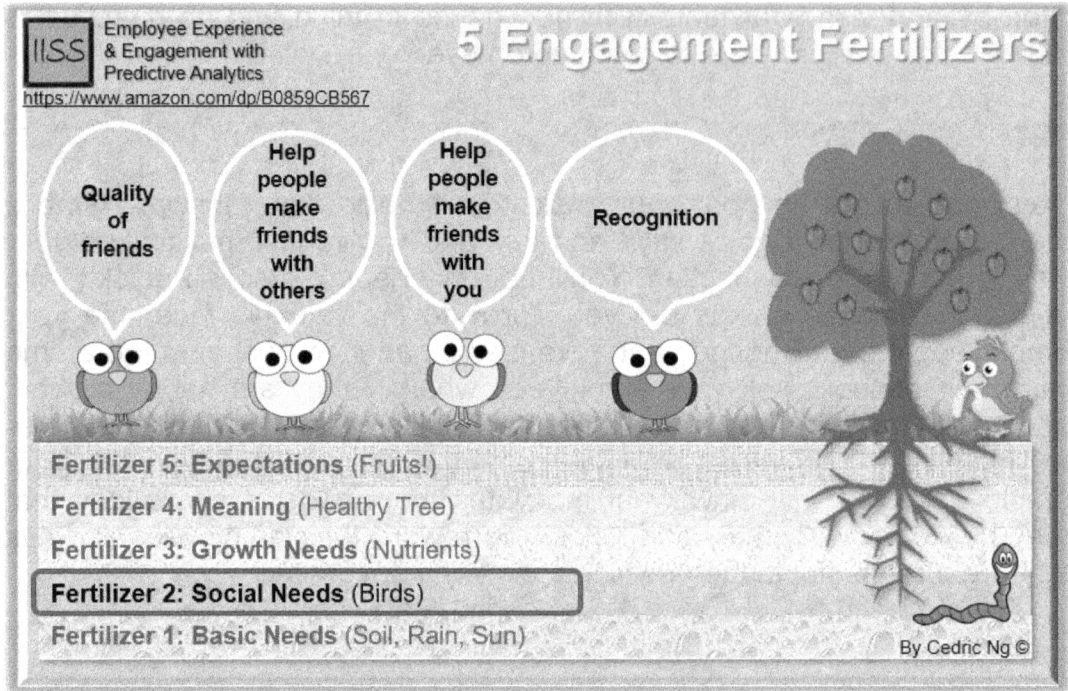

According to research by Deloitte, organizations with recognition programs are highly effective at improving employee engagement & have 31% lower voluntary turnover than their peers with ineffective recognition programs".[1] When people know that their efforts are recognized, you'll soon see a snowball effect.

- **Provide recognitions options**. E.g. Thank you email templates, ePraise cards, Recognition App.

- **Share recognitions ideas with employees**. E.g. Acknowledge your colleagues with a handwritten note, recognize your colleague's hard work during Townhalls.

- **Announce employee achievements** via various channels. E.g. company intranet, departmental meetings, social events.

- **Consider the person's personality when giving recognition**. When deciding how to recognize individuals, take their personalities into account. An extrovert may get a kick out of a public display, but a more introverted person might cringe at such a spectacle. Tailor your approach to what you know of their preferences – and if you aren't sure, ask. [2]

How to give non-cash awards?

Non-cash rewards motivate employees as much as cash bonuses. For non-cash awards to engage employees, follow these three steps: [3]

- **Thank the Employee** — Express your gratitude; it is key to recognize and appreciate your employees.
- **Describe What the Employee Did** — Ensure the employee understands why they are being recognized so that they are more likely to repeat the behavior in the future.
- **Explain How the Employee's Action Added Value** — Describe how the employee's action helped you, the team, or the organization.

Reference
(1) Sareen Babu Madupu (2019) Top 9 Actionable Employee Engagement Ideas and Activities https://acuvate.com/blog/top-9-actionable-employee-engagement-ideas-and-activities/ (3 October 2019)
(2) HBR (2019) It takes more than just extrinsic rewards to inspire people. HBR Guide to Motivating People. United States of America. Harvard Business School Publishing Corporation.
(3) Engagement @ Harvard (2009) Engagement Toolkit for Managers and Leaders http://hr.harvard.edu/files/humanresources/files/engagement_toolkit_leaders_managers.pdf (4 October 2019

3.4 Engagement Fertilizer 3: Growth Needs

Engagement Fertilizers are needed to create the fertile soil for great employee experience & engagement. You will get bigger apples, if you provide your apple tree with fertilizers. Similarly, you will get more engaged employees, if you provide your employees with developmental needs. This is where engagement fertilizer 3 (Growth Needs) comes in. We always want to feel like we're making progress. If we have a craft and we know that we're getting better and better over time, it gives a sense of mastery.

Employee perception of internal opportunities for growth is one of the more important predictors of employee engagement. According to a survey by Glassdoor and Harris Interactive, lack of career growth is the number one reason employees leave a company - only 8% left because of their managers. The same survey also found that more applicants (52%) wanted to hear about growth opportunities when interviewing for a job than about any other perk. [1]

According to Gallup, millennials care deeply about their development when looking for jobs. 87% of millennials rate "professional or career growth and development opportunities" as important to them in a job. [2]

Investing in staff training not only makes them more effective, it also makes them feel valued and more engaged. If their employer doesn't support their development, then they are likely to look elsewhere for one that will.

What is Growth?

Growth – feeling that we are learning, improving and progressing is a universal human need. People become bored and disengaged when they feel that their work is mundane. This need for mastering new skills, accomplishment, and even exceling in high-stress situations, is also known as the need for **achievement**. People want and need challenge. People don't you get excited to go to work because they know that they will have nothing to do. They get excited to work on something that stretches and challenges them.

A dedicate balance

Fertilizers are spurs growth if taken in the appropriate dosage. Similarly, more responsibility can be a source of growth, but it's a delicate balance. A carefully calibrated amount of challenge can energize employees - but too much can overwhelm. Some people worry that if they do their jobs too well, they will be rewarded with more work! For the organizational leader, the challenge of growth is using the right amount of challenge, where each employee feels energized but not overwhelmed by the opportunity to learn new things and develop new skills.

Growth isn't always about advancement.

Organization can create the fertile soil for employee growth through: teaching others, gamification, mobile learning, motivational talks, orientation & onboarding, job rotation, mentoring & coaching.

References
(1) Tracy Maylett, ED.D. (2016) Do Your Employees Grow or Go? https://decision-wise.com/do-your-employees-grow-or-go/ *(11 January 2020)*
(2) Amy Adkins and Brandon Rigoni (2016) Millennials Want Jobs to Be Development Opportunities https://www.gallup.com/workplace/236438/millennials-jobs-development-opportunities.aspx *(3 October 2019)*

3.4.1 Teach others

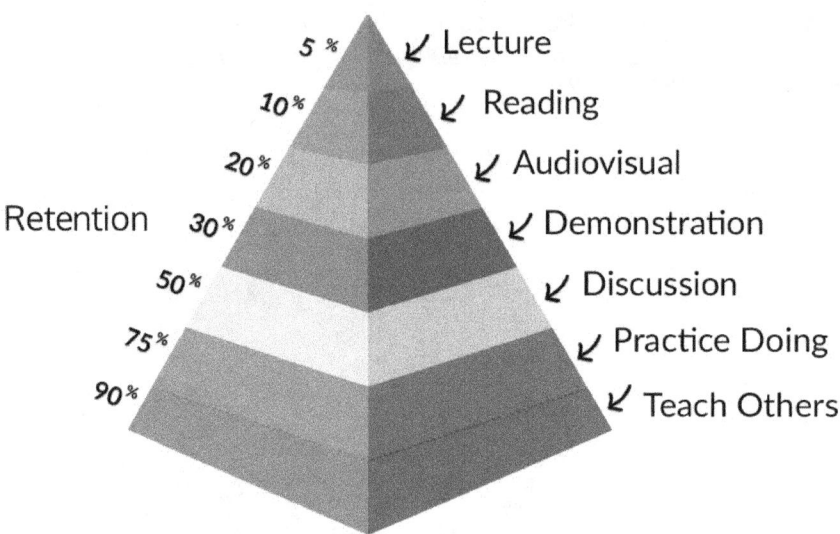

Source: https://medium.com/accelerated-intelligence/explanation-effect-why-you-should-always-teach-what-you-learn-9800983a0ea1

Have you ever felt that you knew something, but then got tongue tied when you tried to explain it to someone else? It happens because teaching reveals gaps in our knowledge. When you learn with the intention to teach you learn more deeply. If you want to become a celebrity teacher, get a flood of fans and customers, teach what you love full time. [1] Once a month, let employees host a lunch & learn on any subject they're passionate about.

In a Harvard study, "employees who spent the last 15 minutes of each day of their training period writing and reflecting on what they had learned did 23% better in the final training test than other employees." [2]

1) Before you teach: When you learn with the intention to teach you learn more deeply
2) While you teach: Teaching reveals gaps in your knowledge.
3) After you teach: Teaching give you valuable feedback from others.

References
(1) Michael Simmons (2019) These Studies Show The Incredible Power Of Teaching To Learn https://medium.com/@michaeldsimmons/these-studies-show-the-incredible-power-of-teaching-to-learn-8b023ac4e556 (3 October 2019)
(2) Michael Simmons (2019) Explanation Effect: Why You Should Always Teach What You Learn https://medium.com/accelerated-intelligence/explanation-effect-why-you-should-always-teach-what-you-learn-9800983a0ea1 (3 October 2019)

3.4.2 Train Managers to Lead

Organizations often promote doers into managerial roles and expect that they will know how to lead. In assessing manager effectiveness, it's generally thought that those who score well on traditional annual evaluations will also score well on the 360-degree feedback assessment - However, studies have found that there was no correlation. Managers can hit their operational metrics and get results, but flop in their 360-degree assessments - This means that they were getting results at the expense of customer and employee relationships. In short, these managers were hitting targets but leaving a trail of bodies as they did so. That is usually the result of putting managers are put into positions where they are expected to lead without knowing how to do so. [1]

References
(1) Tracy Maylett, ED.D. & Paul Warner, PHD (2014) Magic. Five keys to unlock the power of employee engagement (11 January 2020)

3.4.3 Gamify your workplace

Gamification means applying game mechanics to certain activities to drive real-world behaviour change. People naturally want to achieve and compete, so gamification combines personal motivations with company goals. Game mechanics can easily be applied in the workplace in a number of ways. As gamification makes day-to-day activities like training more fun, using it can boost eLearning courses participation rates and increase employee motivation. Firms can award achievement levels and badges for quick completion or top scores, which can be published to all employees. So instead of simply getting it done, employees will want to win, which means finishing correctly and quickly. [1]

Gen Y grew up playing video games. An MTV report shows that 74 percent of Gen Y agree that if the workplace were like a game, they know how to level-up faster than others. According to Forbes, at Bluewolf Consulting, employees earn points by posting creative new topics for discussion or responding to posts of others. Points can then be traded in for prizes including lunch with the CEO and an iPad. As a result, website traffic has increased by 80 percent. Gartner predicts that over half of organizations that manage innovation processes will gamify them in the next three years. [2]

References
(1) Lamont Exeter (2019) Three Ways to Digitize the Workplace to Boost Employee Engagement https://www.ttec.com/articles/three-ways-digitize-workplace-boost-employee-engagement (3 October 2019)
(2) Dan Schawbel (2012) 5 Ways to Retain Gen Y Workers https://www.americanexpress.com/en-us/business/trends-and-insights/articles/5-ways-to-retain-gen-y-workers/?fbclid=IwAR0a2431isZHKS5JPeZGmU3GVXsp4pz5qX264XeD4CpdxoV5WfqL7t3Fvl4 (12 March 2020)

3.4.4 Mobile learning

Mobile technology is one of the best ways you can create better employee engagement. In 2014, Bersin released it's Global Human Capital Trends Survey, which found that mobile learning is changing employee training. Mobile learning and online learning let learners access content and complete training at the time, pace, and place that suits them. Mobile learning enables flexible working practices and raise enthusiasm towards learning. People are happier to complete learning on their mobile at home or wherever they might be. [1]

References
(1) Admin (2018) *5 Secrets to Increase Employee Engagement With Technology* https://www.shiftelearning.com/blog/increase-employee-engagement-with-technology (3 October 2019)

3.4.5 Motivational talks

Start Monday off with some inspiration, by
- finding an inspirational quote and sending it to your team to motivate and inspire them.
- hiring a motivational speaker to come into the office. Bringing in a motivational speaker can work wonders for engagement. They may have an inspirational story to share or motivational techniques that drive greater team spirit. [1]

References
(1) Rever Team (2019) 15 Employee Engagement Activities For Your Team https://reverscore.com/employee-engagement-activities/ (2 October 2019)

3.4.6 Orientation & onboarding

Employees should be engaged from their very first day. Orientation and onboarding are different types of employee trainings. While orientation training is a one-time session with a checklist-style agenda delivered to everyone in the company, onboarding training are department-specific sessions that take place over a longer period of time. Onboarding trainings ideally start on the first day of employment and may carry on throughout the first year as needed. Effective and successful onboarding trainings bring new employees up to speed much quicker than if left to fend for themselves. Begin by evaluating your current orientation and onboarding process: [1]
- Will it leave new hires feeling inspired or demotivated?
- Will it help new hires build connections with other team members?

With up to 20% of staff turnover occurring within the first 45 days of employment, standardized onboarding process is essential. Organizations with standardized onboarding process experience 62% greater new hire productivity with 50% greater new hire retention. There are three dimensions of onboarding (Organizational, Technical, Social): [2]

1) Organizational Onboarding

Help new hires assimilate & teach them how things work. Organizational Onboarding should cover the following: Company mission, vision, values, culture, organizational structure, leadership team, key corporate policies, staff benefit plans, glossaries of terms, administrative procedures (computer logins, telephone extension, email setups, etc.)

2) Technical Onboarding

Just because someone is hired for their capabilities and experiences, doesn't mean they know how to deploy them at your company. Schedule weekly coaching sessions to check in and ensure they have opportunities to make meaningful contributions as soon as possible. Help new hires set up early wins - Start with targets you are confident your new hires can meet. Gradually increase the level of responsibility associated with each task.

3) Social Onboarding

New hires who feel like a stranger in a foreign land have a high chance of leaving a job. New hires, in partnership with their manager, should identify 7-10 people (superiors, peers, direct reports, and internal and external customers) whose success they will contribute to, or who will contribute to their success. The new hire should then craft plans to connect with each stakeholder, one-on-one. This can be a short meeting over coffee or lunch — an opportunity to learn and ask for guidance. In addition to stakeholder cultivation, building social capital with teammates on a daily basis helps build camaraderie.

References
(1) Rever Team (2019) 15 Employee Engagement Activities For Your Team https://reverscore.com/employee-engagement-activities/ (2 October 2019)
(2) Ron Carucci (2018) To Retain New Hires, Spend More Time Onboarding Them https://hbr.org/2018/12/to-retain-new-hires-spend-more-time-onboarding-them?utm_source=facebook&utm_medium=social&utm_campaign=hbr&fbclid=IwAR2Nv6YwxqQqg0ij6xTuZUZ6IlQ70rDikPH-rPaeO--HNtf0PXhwCXEGYi8 (4 October 2019)

3.4.7 Job rotation

Job rotation is one of the most effective employee engagement activities. Letting people job rotate to work in other teams, rejuvenates them, and helps them to learn new skills. It enables them to see other perspectives, support cross-functional collaboration, and facilitate new ideas. Job rotation also give employees something to aspire to and aim for. So, even if you can't offer them to everyone now, it may inspire staff to work towards it. [1]

Employees don't always have to change companies to get the development they want. Implementing job rotation at the global, regional, and local level, can help you retain talent. Thus, global organisations often have job rotation programs for their high-potential employees. In Job rotation, employees rotate between jobs within the organization. They take on new tasks at a different job for a period of time before rotating back to their original position. Job rotations promote flexibility, employee development, engagement, and retention.

Benefits of job rotation

1. Develops and retains employees
Employees who work at a company that encourages their development might not feel the need to change jobs. Instead of leaving your organisation for a new job, they can rotate jobs.

2. Helps you identify where employees work best
You might find that an employee can better handle a different job at your organization. For organizational effectiveness, you need to have all your employees in the right positions.

3. Backup plan if an employee leave
By having a job rotation plan, you have several employees who know how to do each job, you won't need to scramble to hire the first person you see, when an employee leave, as you have other employees capable of covering the separated employee's tasks. If you need to hire a replacement, you can take your time to find the right fit.

Issues of job rotation

1. Employees may become disgruntled employees
Some employees might not want to rotate jobs. You might have some employees who excel at their job but aren't willing to learn new things.

2. Might not be feasible for some countries and positions
For some countries and positions, job rotation is not possible. This is especially true in highly skilled positions where employees need years of training to do their jobs.
- **Indonesia**: In Indonesia, foreigners cannot work in functions such as Legal, Supply chain management, Human resources, Quality inspection and control, Environment affairs, Health and safety
- **Saudi Arabia:** Expatriates are banned from all jobs relating to Human Resources as well as positions in the departments dealing with hiring workers in the private section.

3. Your business may suffer
When you move an employee to another country or to a new position, there is a learning curve. Because employees are learning new skills, there could be errors. Customers could become frustrated by confused employees who make mistakes. If operations don't run smoothly, your bottom line could suffer.

How to implement job rotation

1. **Get leadership support**
   Get leadership support with a business case. E.g. With job rotation you are creating an internal talent pool to minimise the impact of turnover for critical positions.

2. **Determine the critical positions for job rotation.**
   Assess the future growth plans of the organization, and its vulnerabilities in terms of each position's average tenure, turnover risk for each role, and list of employees who are retiring soon,

3. **Create job readiness assessments.**
   Develop assessments to determine candidate readiness and current competencies, and tailor the job rotation experience.

4. **Determine readiness periods.**
   A readiness period is an estimate of how long a job rotation program should be. There are different readiness periods for different roles.

5. **Develop the selection process.**
   Determine how will candidates be selected. In the US, if the job rotation program is used for promotions, employers need to ensure that there is no discrimination in the selection process, as per Equal Employment Opportunity Commission (EEOC) guidelines.

6. **Support the process.**
   Prepare the people who are going to be working with the individual, including the individual themselves. Help them to understand what their roles are in the job rotation program and how to do it. Use check-in milestones along the way to ensure that progress is being made against the program goals. Prepare the team for the job rotation via online learning, guidelines, mentoring, buddy system, etc.

References
(1) Rever Team (2019) 15 Employee Engagement Activities For Your Team https://reverscore.com/employee-engagement-activities/ (2 October 2019)

3.4.8 Coaching

Be a Coaching Manager - Managers needs to be a coach to their employees to bring out their best. The GROW Model is a simple framework for structuring coaching conversations. It helps employees to establish a goal, examine the current reality, explore options, and decide what they will do. [1]

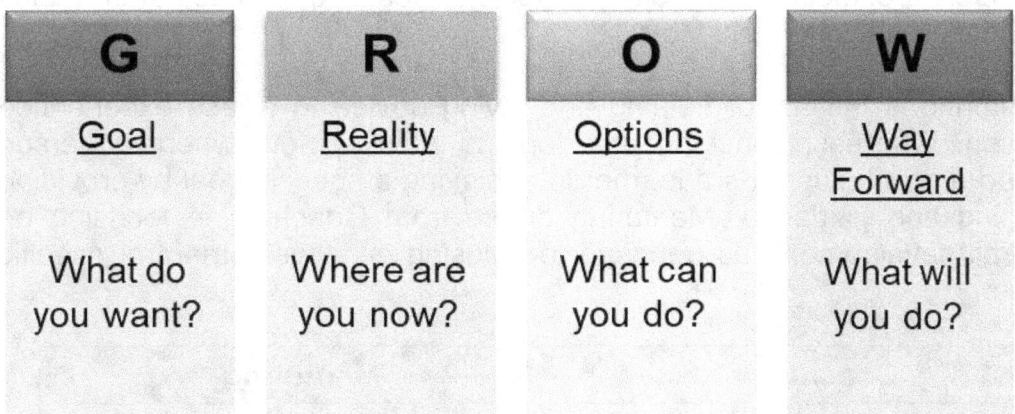

The GROW Model is a simple yet powerful method for goal setting and problem solving. It is commonly used to structure coaching and mentoring sessions.
- **Goal**: Goal is the end point, where the mentee or coachee wants to be. It is important to set goals that are SMART (Specific, Measurable, Attainable, Realistic, Time-bound).
- **Reality**: Reality is where the mentee or coachee is now. What are the issues, the challenges, how far are they away from their goal?
- **Options**: What are the options to achieve the goal.
- **Way Forward**: Steps that will take the mentee or coachee to their goal.

Reference
(1) Engagement @ Harvard (2009) Engagement Toolkit for Managers and Leaders
http://hr.harvard.edu/files/humanresources/files/engagement_toolkit_leaders_managers.pdf (4 October 2019

3.4.9 Mentoring

A study done by Deloitte in 2012 found that retention is 25% higher for employees who have engaged in company-sponsored mentorship. [1]

Helping others is one of the ways to build satisfaction. Create mentorship opportunities by setting up a mentoring program. Mentoring makes team members feel more engaged and helps your less experienced staff to they reach their full potential faster.

Mentoring is the act of helping and giving advice to a less experienced person. Whereas, coaching is a form of development where a person called a coach supports a learner in achieving a specific goal by providing training and guidance. Mentoring differs from Coaching in focusing on overall development, as opposed to focusing on development of specific skills.

Coaching	Mentoring
Focus on development of specific skills	Focus on overall development
Focus on telling	Focus on listening
Shows you where you went wrong	Helps you work it out for yourself
Open	Confidential
Coach may line own success to that of coachee	Mentor does not link own success to that of mentee
Encourage self-improvement	

IBM has been a pioneer of personalized employee development. One approach they've used to support people is a mentorship matching program. Rather than creating a one-size fits all employee development system, IBM provides an online database that allows employees to find a mentorship relationship that can help them fulfill their development objectives. There are no assumptions around who should mentor whom, and mentor relationships exist across business units and geographical locations. (2)

References
(1) Amy Adkins and Brandon Rigoni (2016) Millennials Want Jobs to Be Development Opportunities https://www.gallup.com/workplace/236438/millennials-jobs-development-opportunities.aspx (3 October 2019)
(2) Natalie Baumgartner, HBR (2019) One engagement strategy does not fit all. HBR Guide to Motivating People. United States of America. Harvard Business School Publishing Corporation.

3.4.9.1 Benefits of having a structured mentoring program

According to the Association for Talent Development, more than 70 percent of Fortune 500 companies have some type of mentorship program. Employees can turn to their mentor for career guidance, and to learn about company inner workings. Mentoring is an impactful strategy to develop, engage and retain your people. Some of the benefits of having a structured workplace mentoring program at your company are as follows:

1. Reduces cost of learning

Cost of internal workplace mentoring programs is minimal.

2. Creates a learning culture

A workplace mentoring program creates a learning and collaborative culture. Employees know their future is being invested in if they're assigned a mentor who can help them in their career development. When a new employee who's assigned a mentor benefits from that relationship, the mentee may be more likely to volunteer to do the same for other employees. By creating a formal program, employees will find it easier to join, follow guidelines and make a difference.

3. Reduces stress and anxiety

When faced with an issue at work, employees may not approach their manager or teammates. A workplace mentor is another person that they can turn to for guidance.

4. Increases retention, promotion rates and employee engagement

Mentoring increases retention, promotion rates, and employee satisfaction. According to a study published in the Plastic Surgical Nursing journal, the mentor experience/relationship positively influenced job satisfaction of new hire nurse practitioners. And a higher level of satisfaction is associated with reduced turnover and better patient experience.

5. Benefits both the mentees and mentors

Mentoring benefits both the mentees and mentors. Its drives learning and development for both mentees and mentors.

- **Mentees**: Helps the mentee better understand the organization's culture and unspoken rules, and provides an important networking contact for the mentee.

- **Mentor:** Opportunity for the mentor to learn about other functions, develops the mentor's coaching and listening skills. According to a 2013 study, "Career Benefits Associated with Mentoring for Mentors," published in the Journal of Vocational Behaviour, mentors versus non-mentors were more satisfied with their jobs and had a stronger sense of commitment to the organization as a whole.

3.4.9.2 How to setup a mentoring program?

Great mentoring programs don't just happen. Here's a five-step process to create a high-impact mentoring program.

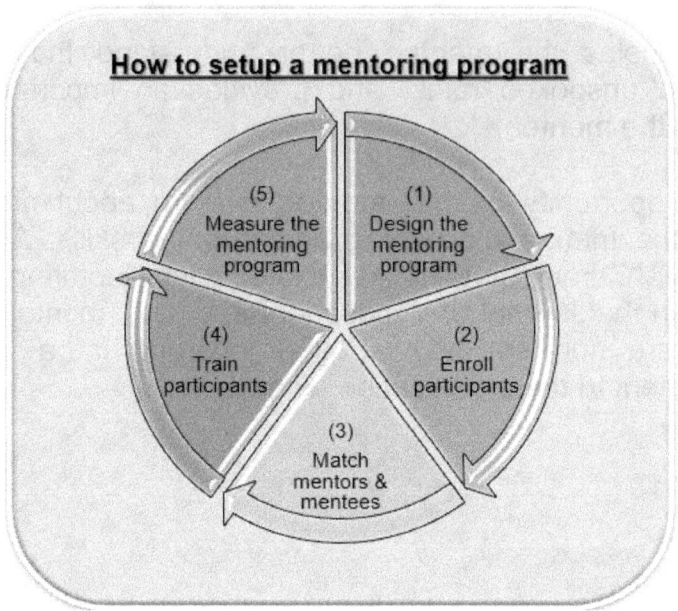

Step 1: Design the mentoring program

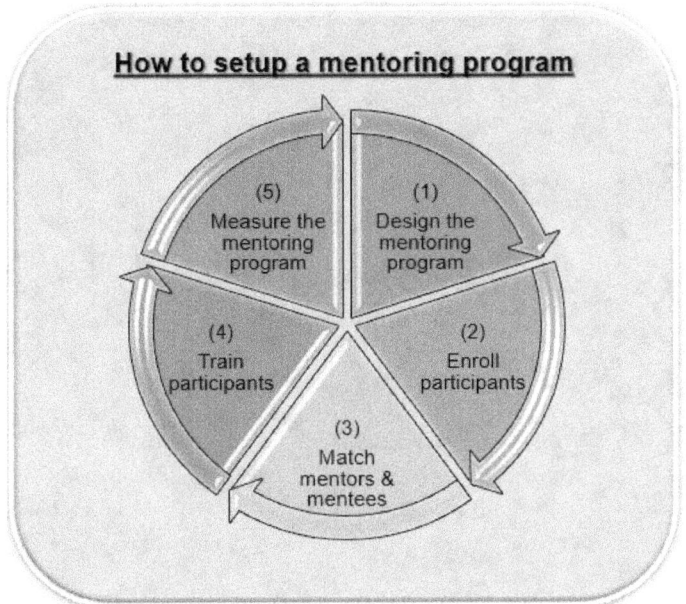

Effective mentoring programs is structured and flexible. There should be a program workflow diagram to explain each step of your program with details such as enrolment, duration, etc.
- **Structure** provides participants a mentoring workflow to follow.
- **Flexibility** support varying individual mentoring needs depending on their learning goals, preferences, and learning style.

Components of a mentoring program design include:
- **Enroll participants:** is the mentoring program open to all employees, or invite only?
- **Match mentors and mentees**
- **Train participants**
- **Measure the mentoring program**

Step 2: Enroll participants

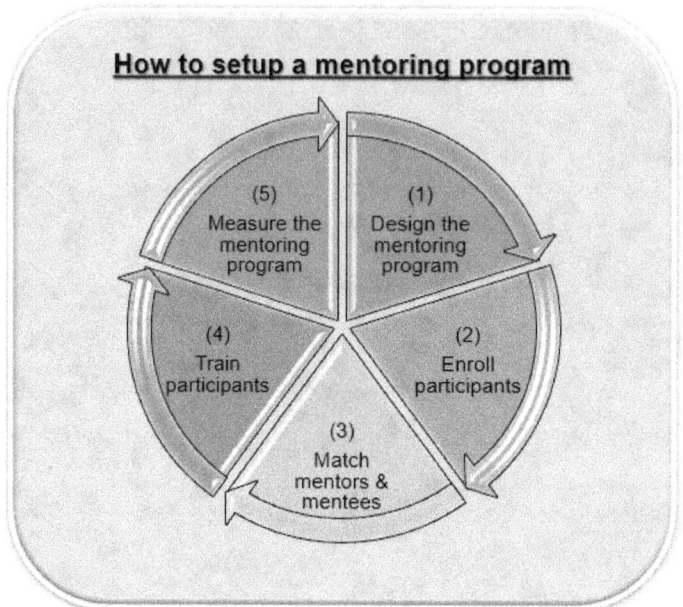

Decide if the mentoring program open to all employees, or invite only? Mentoring programs won't succeed without effective promotion, mentor recruitment, and training.

- **Promote the benefits of mentoring.**
  Building a pool of mentors can be a challenge. Mentors are usually busy people. Educate participants, leaders and stakeholders on the benefits of the mentoring program and value to the organization.

- **Offer rewards and recognition participation.**
  Formally recognizing mentor involvement can be motivating and help attract additional mentors to the program.

Step 3: Match mentors and mentees

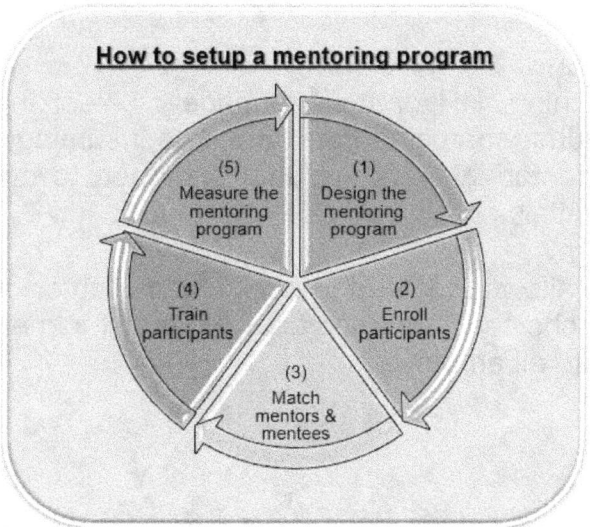

A productive mentoring relationship depends on a good match. Matching is one of the most challenging aspects of a mentoring program, as participants have different backgrounds, competencies, learning styles and needs. Steps to successful mentor matching:

1. Create participant profiles.
Create a profile for all participants (mentors and mentees). The information about the participants, the better chance the participants can get a better mentoring fit. Participants profile should include:
- Development goals and interest,
- Job function,
- Education and experience,
- Location,
- Matching preferences.

2. **Decide on the mentoring matching method.**
Matching starts by deciding which type of matching you'll offer in your program.
- **Admin-matching.**
  In admin-matching, HR or their managers select the mentor for the mentee. As admin-matching can be time-consuming, this approach is suitable for mentoring programs that are open to only a few people. e.g. high-potentials only.
- **Self-matching.**
  Self-matching allows mentees to select their mentor or submit their top three mentor choices. Self-matching is ideal for mentoring programs that are open to all employees.

Step 4: Train participants

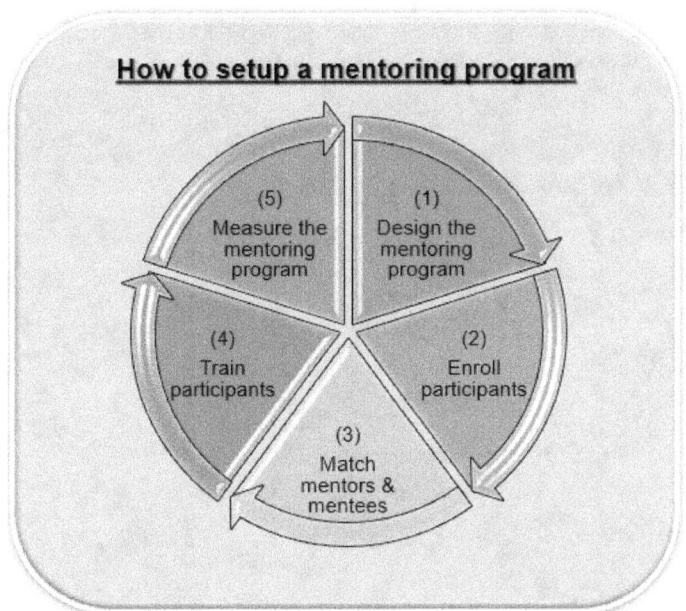

Productive mentoring doesn't just happen. Now that your participants are enrolled and matched, you need to providing some guidelines and training to mentors and mentees.

- **Participant roles.**

Educate the mentor and mentee about their roles in the mentorship.

- **Goals and action plans.**

Ensure all mentorships have goals and action plans. This brings focus, and adds accountability to accomplish something.

- **Mentoring duration:**

Decide on the mentoring duration. Is it just a single session, 5 weeks, 6 months, or 1 year?

- **Mentoring best practices.**

Provide mentoring tips and best practices throughout the mentoring program to help participants stay on track and get the most out of the program.

Step 5: Measure the mentoring program

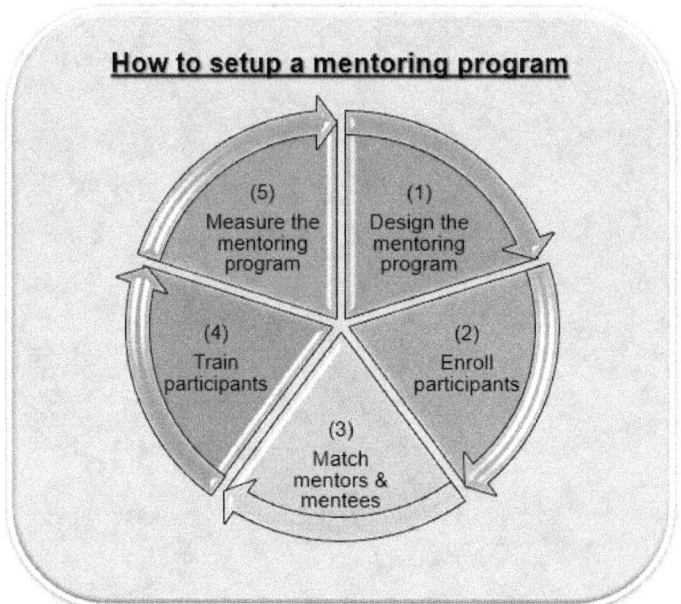

Measuring the impact of mentoring is important to get ongoing support. The measure phase is essential to identify trouble spots and opportunities. Even if your organization doesn't want to formally track the details, just the act of reporting progress helps mentors and mentees stay productive.

Mentoring programs should be tracked, measured, and assessed.
- **Measure program effectiveness**

For example, you can compare promotion rates of program participants to non-participants.
- **Measure mentee development**

One of the easiest ways to capture outcome and feedback is through surveys. Ask participants and stakeholders how well the mentoring program met their goals and the goals of the organization.

3.4.9.3 Guidelines for mentors

Becoming a mentor is worth your time. You get to share your experience and learn about challenges you might not be aware of. A mentee can be your eyes and ears at the front lines. If you are a first-time mentor, there are a few things you need to know to make the mentoring experience worthwhile on both sides of the table.

1. Help your mentee define their goals

This is your first task as a mentor. Help your mentee define their goals, then structure your conversations around them. Make sure that the goals are SMART (Specific, Measurable, Achievable, Realistic, Time-bound). A good rule of thumb is to focus on one goal over three months, and check in regularly to measure progress and adjust as you go.

2. Understand the context

Before you dive into problem solving, spend some time to understand the context. Context can help you better understand their goals, the type of help they've received in the past, their past efforts, and how you're best suited to help them make progress.
- What are they grappling with most right now?
- What's their relationship like with their manager and colleagues?
- How do they think you can help?

3. Commit to a meeting schedule

You may wish to meet once every two weeks for the first couple of sessions, but after that you may change to once per month, or whatever is appropriate for you both. If you can't commit to meeting regularly, now isn't the right time for you to be a mentor.

4. Keep an open mind to problem solving

Be open to the possibility that your advice might not always be right for them, and let them know you're open to their feedback.

5. Make Referrals

Whenever your mentee needs help beyond your expertise, connect them with people who are more knowledgeable than you. You won't have the answer to every question, but that doesn't make you a lousy mentor. Great mentors recognize when there's a more relevant resource at hand – whether that's an article, book, or person in their network.

6. Maintain confidentiality

What your mentee tells you should be kept confidential - that's the basis of your mentoring relationship, and that is why your mentee comes to you rather than their manager. However, if your mentee is facing issues such as sexual harassment, then it's best to loop in HR.

7. Practice active listening and questioning

Active listening takes practice. You may find yourself mentally problem solving for your mentee while they are still talking through a problem. Ask questions to clarify their views and encourage their own thinking on the challenge. Effective listening does not happen by chance - it takes a lot of practice to develop. Our ability to listen effectively gives us an advantage in our life. There are **three levels of listening**:

Level 1: Internal Listening
This is the lowest level of listening and comes naturally, unless we intentionally develop our listening skills. At this level, you are not really listening to others when they talk, instead, you are thinking about the next thing to say, while they are talking. Listening to speak often create misunderstandings and causes us to miss key information.

Level 2: Focused Listening
At this level we are actively paying attention to what the other person is saying. We are not thinking about what we want to say next or distracted by other things.

Level 3: Global Listening
This is the highest level of listening and you will need to practice to get there. At this level you are not only paying attention to what others are saying, but also what they mean. People say things all the time but often fail to convey the underlying feelings behind their words.

3.5 Engagement Fertilizer 4: Meaning

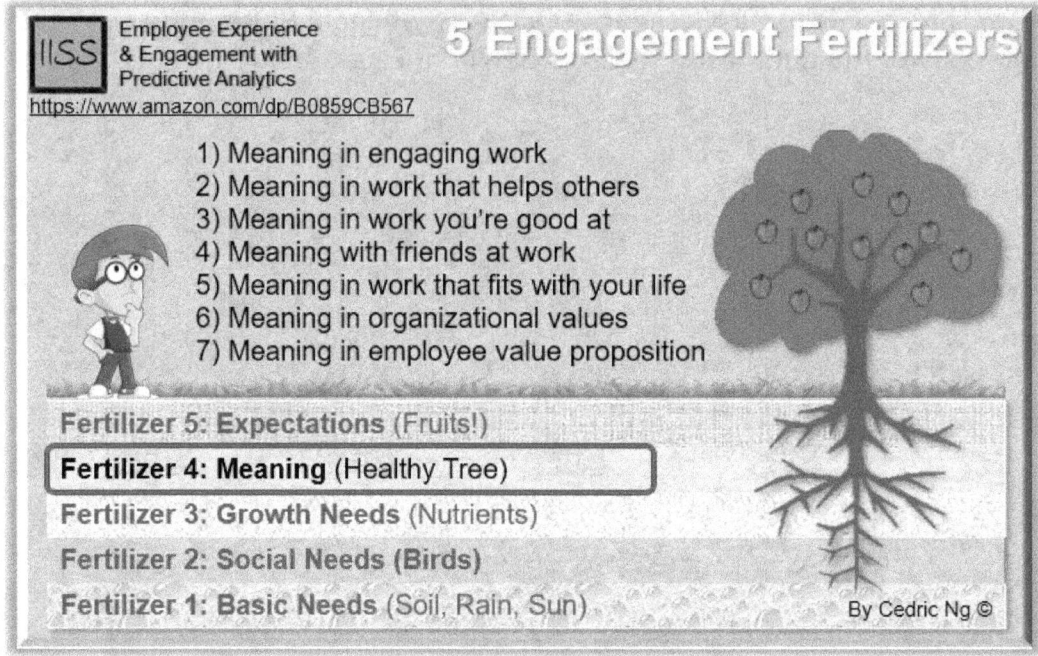

Engagement Fertilizers are needed to create the fertile soil for great employee experience & engagement. For a Tree to be healthy, you need to provide the right type of fertilizers. Using the wrong type of fertilizers can kill the tree. This is where engagement fertilizer 4 (Meaning) comes in. Employees working in the "coolest places to work on the planet" can be disengaged if they cannot find meaning in their work.

We all want to find a dream job that's enjoyable and meaningful, but what does that actually mean? Some people think that the key elements of their dream job are that it be easy and highly paid. Meaningful jobs aren't as simple as "following your passion". Steve Jobs was passionate about Zen Buddhism before entering technology. Rather, you can develop passion by doing work that you find enjoyable and meaningful.

Organizations can till the soil and lay down the fertilizers that allow people to create their own meaning out of work through:
- Meaning in the work itself
- Meaning in work that helps others
- Meaning in work you are good at
- Meaning with supportive colleagues
- Meaning in work that fits with your life
- Meaning in values
- Meaning in Employee Value Proposition (EVP)

What is Meaning?

You've heard people will work for money, but die for a cause.

Meaning relates to question: Why are we here? People at all levels need to know that they are much more than just cogs in a machine which they have no control. A cog in the machine describe someone who has only a small role in an organization - someone who is insignificant. People need to know that what they are doing is having an impact. In reality, if a cog does not work correctly in a machine, the entire machine will not work - a person's role in a project may be small, but it is important.

Money makes you happier, but only a little

A survey by D. Kahneman and A. Deaton, found that money does make you happy, but only a little. People were asked to rate how satisfied they were with their lives on a scale from one to ten. The result is shown on the right, while the bottom shows their household income. These lines are almost flat after $75,000, so beyond this point, income had little relationship with how satisfied people felt. This is hardly surprising – we all know people who've gone into high earning jobs and ended up miserable.
(1)

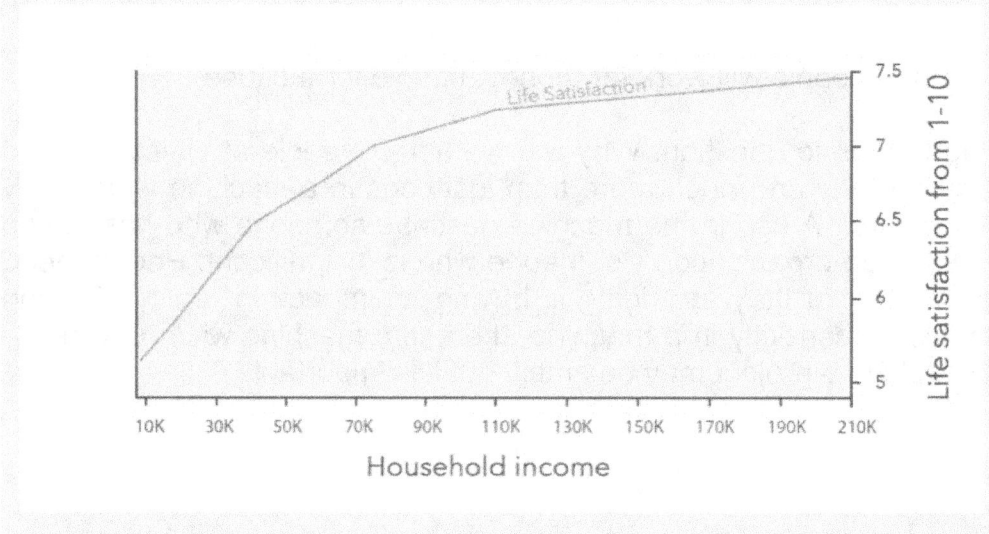

High income improves evaluation of life but not emotional well-being, D. Kahneman and A. Deaton, 2010

Source: https://80000hours.org/career-guide/job-satisfaction/#3-work-youre-good-at

The organization can lavish you with perks and development needs, but they may not hold the key to engagement. People need meaning in life and in work. Think about a time when you received a big salary raise. That raise fueled your motivation and you were ready to slog long hours in appreciation of your boss generosity. But that motivation only lasted a short time, then things went back to normal. It is not monetary, but meaningful work that has lasting power. Frederick Herzberg, the father of motivational theory, said, "I can charge a person's battery, and then recharge it and recharge it again. But it is only when one has a generator of one's own that we can talk about motivation. Shaping one's work involves finding those intrinsic motivators that, like the generator, keep people engaged in what they do. (2)

Salary, perks, and creating shareholder value don't give people a sense of mission and purpose – they are the basic requirements for engagement. People create their own meaning. When meaning is present, even mundane, repetitive work becomes much more than that. The opposite is also true - Employees working in seemingly challenging and fulfilling jobs can be disengaged if they cannot find meaning in them. Money may lure people into jobs, but meaning, and the prospect of interesting and valuable work determine both their tenure and how hard they will work while they are on the job. If you ask a coworker the question, "What does a good day look like at work?", their response seldom relates to compensation and perks. People usually relate a good day with having accomplished something important. Organizations can till the soil for meaning to grow by having a clear organizational mission, sharing the corporate story, and explaining to everyone "this is why we do what we do".

Reference
(1) Benjamin Todd (2017) We reviewed over 60 studies about what makes for a dream job. Here's what we found https://80000hours.org/career-guide/job-satisfaction/#1-work-thats-engaging (4 October 2019)
(2) Timothy R. Clark (2012) The Employee Engagement Mindset (11 January 2020)

Stress is bad if you think it is

Many people assume stress is obviously bad, and lots of people tell us they want to find a "low stress job". But according to psychologist Kelly McGonigal stress is only bad if you think it is, and that stress can make us stronger, smarter and happier. In summary, whether work demands have good or bad effects seems to depend on the following things: [1]

Variable		Good (or neutral)	Bad
Type of stress	Intensity of demands	Challenging but achievable	Mismatched with ability (either too high or too low)
	Duration	Short-term	On-going
Context	Control	High control and autonomy	Low control and autonomy
	Power	High power	Low power
	Social Support	Good social support	Social isolation
How to cope	Mindset	Reframe demands as opportunities, stress as useful	View demands as threats, stress as harmful to health
	Altruism	Performing altruistic acts	Focusing on yourself

Source: https://80000hours.org/2016/02/should-you-look-for-a-low-stress-job/

<u>Reference</u>
(1) Roman Duda (2016) Will high stress kill you, save your life, or neither? https://80000hours.org/2016/02/should-you-look-for-a-low-stress-job/ (25 February 2020)

Should you follow your passion?

When we look at successful people, they are often passionate about what they do. So, should you just follow your passion, and pursue a career involving that interest?

"Follow your passion" can be misleading advice. You can become passionate and good in new areas, if you practice, if you work on engaging tasks, if your work helps others, if you work with people you like. Many successful people are passionate, but often their passion developed alongside their success, and took a long time to discover. Steve Jobs started out passionate about Zen Buddhism. He got into technology as a way to make some quick cash. But as he became successful, his passion grew, until he became the most famous advocate of "doing what you love". In reality, rather than having a single passion, our interests change often. Think back to what you were most interested in 10 years ago, and you'll probably find that it's pretty different from what you're interested in today. [1]

<u>Reference</u>
(1) Benjamin Todd (2017) *We reviewed over 60 studies about what makes for a dream job. Here's what we found* <u>https://80000hours.org/career-guide/job-satisfaction/#1-work-thats-engaging</u> *(4 October 2019)*

Meaningful work by industries and jobs

Technology
In 1983, Steve Jobs was trying to entice John Sculley to leave PepsiCo to become Apple's new CEO. Jobs asked him, "Do you want to spend the rest of your life selling sugared water or do you want a chance to change the world?" In making his pitch, Jobs leveraged a potent psychological force: the deep-seated human desire to do meaningful work. [1]

Cleaner
A cleaner at NASA who, when asked by John F. Kennedy what his job was, responded, "I'm helping to put a man on the moon." This anecdote is often used to show how even the most mundane job can be seen as meaningful with the right mindset and under a good leadership. [2]

Restaurant
Restaurant-owner, Ari Weinzweig who leads a multi-million dollar enterprise, still pours water to guests, every single evening - It's when employees see Ari walking around with a pitcher offering water to clients they know their own jobs are equally important. [3]

Lifeguards
Adam Grant of the Wharton School found that lifeguards were more vigilant after reading stories about people whose lives have been saved by lifeguards. Harvard Business School's Ryan Buell found that when cooks see those who will be eating their food, they feel more motivated and work harder. [4]

Medical devices
Medical device maker, Medtronic's CEO Bill George regularly brought patients in to meet with the company's employees so they would see how their work was making with the company's employees so they would see how their work was making a difference in the lives of real people. [5]

Reference
(1) Teresa Amabile and Steve Kramer, HBR (2019) The Power of Small Wins. HBR Guide to Motivating People. United States of America. Harvard Business School Publishing Corporation.
(2) Lewis Garrad and Thomas Chamorro-Premuzic, HBR (2019) The best leaders show people that their work matters. HBR Guide to Motivating People. United States of America. Harvard Business School Publishing Corporation.
(3) Joost (2011) How Real Leaders Melt The Iceberg of Ignorance With Humility https://corporate-rebels.com/iceberg-of-ignorance/ (6 January 2020)
(4) Joseph Grenny, HBR (2019) Great storytelling connects employees to their work. HBR Guide to Motivating People. United States of America. Harvard Business School Publishing Corporation.
(5) Tracy Maylett, ED.D. & Paul Warner, PHD (2014) Magic. Five keys to unlock the power of employee engagement (11 January 2020)

3.5.1 Meaning in engaging work

"Choose a job you love and you will never have to work a day in your life"
- Confucius

Job Characteristics Model

Job Characteristics Model

5 Job Characteristics:
- Autonomy
- Skill Variety
- Task Identity
- Task Significance
- Feedback

In the "job characteristics model" job satisfaction is largely determined by how engaging the job itself is. According to the theory, five core job characteristics lead to enriching jobs. [1]

- **Skill Variety**: The degree to which a job requires various activities, requiring the employee to develop a variety of skills. Jobholders experience more meaningfulness in jobs that require several different skills and abilities than when the jobs are elementary and routine.
- **Task Identity**: Employees experience more meaningfulness in a job when they are involved in the entire process rather than just being responsible for a part of the work.
- **Task Significance**: Employees experience more meaningfulness in a job that affects other people's lives either in the immediate organization or in the external environment.
- **Autonomy**: Autonomy is the degree to which the job provides the employee with significant discretion to plan out the work and determine the procedures in the job, rather than on the instructions from a manager or a manual of job procedures. With autonomy, jobholders experience greater personal responsibility for their own successes and failures at work.
- **Feedback**: When employees receive specific, actionable feedback about their work performance, they have better overall knowledge of the effect of their work activities, and what specific actions they need to take to improve their productivity.

"Job Characteristics Model" vs "IISS Model's Engagement Fertilizers"

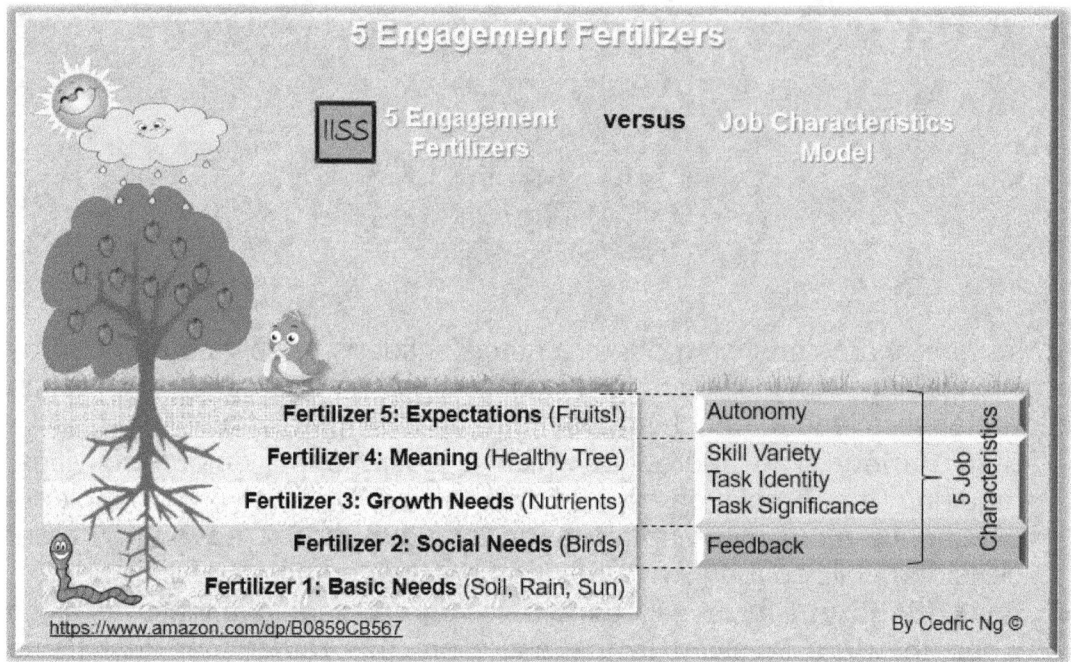

The IISS model's Engagement Fertilizers is similar to the Job Characteristics Model.:
- **Fertilizer 2 (Social Needs)** is similar to Job Characteristics Model's "Feedback".
- **Fertilizer 3 (Growth Needs) and Fertilizer 4 (Meaning)** is similar to Job Characteristics Model's "Skill Variety", "Task Identity", and "Task Significance".
- **Fertilizer 5 (Expectations)** is similar to Job Characteristics Model's "Autonomy"

Reference
(1) Wikipedia (2020) Job characteristic theory https://en.wikipedia.org/wiki/Job_characteristic_theory *(4 October 2019)*

3.5.2 Meaning in work that helps others

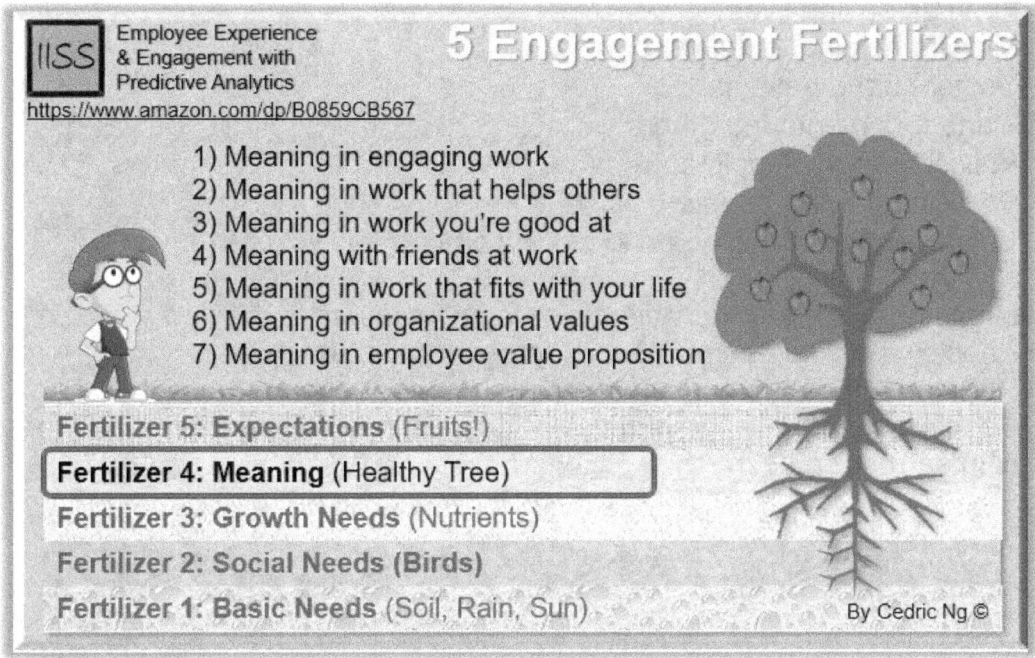

Focus on your community instead of just profit. All businesses should try to make a positive impact on society. If you make it your mission to help others, you'll be more successful, because more people will want to help you succeed, and because they're more motivated by a sense of purpose.

According to Deloitte, Profit as the sole measure of success is rejected by 92 percent of Gen Y and over 50 percent believe that the purpose of business is primarily innovation and societal development. Make sure that your company is giving back to your community and that your employees gets to be part of the process. Gen Y is attracted to non-profit companies because there is a purpose behind them over just getting donations. [1]

These jobs help other people - that's why they are meaningful to most people who does them:
- Fireman saving lives
- Midwife helping to deliver babies
- Doctor curing patients
- Managers mentoring others
- Police keeping people safe
- Politicians giving people a voice
- Volunteers providing a service for a community

Helping others is a key ingredient for life satisfaction.
- People who volunteer are less depressed and healthier.
- Performing a random act of kindness makes the giver happier.
- People who donate to charity are as satisfied with their lives as those who earn twice as much.

References
(1) Dan Schawbel (2012) 5 Ways to Retain Gen Y Workers https://www.americanexpress.com/en-us/business/trends-and-insights/articles/5-ways-to-retain-gen-y-workers/?fbclid=IwAR0a2431isZHKS5JPeZGmU3GVXsp4pz5qX264XeD4CpdxoV5WfqL7t3Fvl4 (12 March 2020)

3.5.3 Meaning in work you're good at

Being good at your job leads to higher job satisfaction, whilst not having the knowledge and skills to do your job well is likely to lead to stress. Most theories of human well-being and needs have achievement as a key component (including Self-Determination Theory). The job characteristics model also has skills as a job characteristic that lead to job satisfaction. [1]

If you do something you are good at that others value, you will have plenty of career opportunities, which gives you the best chance of finding a dream job with all the other meaningful ingredients of work – engaging work, work that helps others, friends at work, fit with your life, values, etc.

Sometimes, skill is more important than passion, when it comes to meaning: [2]
- Being good at your work gives you a sense of achievement, a key ingredient of life satisfaction.
- Even if you love art, if you pursue it as a career but aren't good at it, you'll not be successful. That's not to say you should only do work you're already good at. However, you want the potential to get good at it.
- Being good at your work gives you the power to negotiate for the other components of a fulfilling job, such as the ability to work on meaningful projects, undertake engaging tasks and earning fair pay.

Reference
(1) Roman Duda (2014) Job satisfaction research https://80000hours.org/articles/job-satisfaction-research/#2-work-that-helps-others *(25 February 2020)*
(2) Benjamin Todd (2017) We reviewed over 60 studies about what makes for a dream job. Here's what we found https://80000hours.org/career-guide/job-satisfaction/#1-work-thats-engaging *(4 October 2019)*

3.5.4 Meaning with friends at work

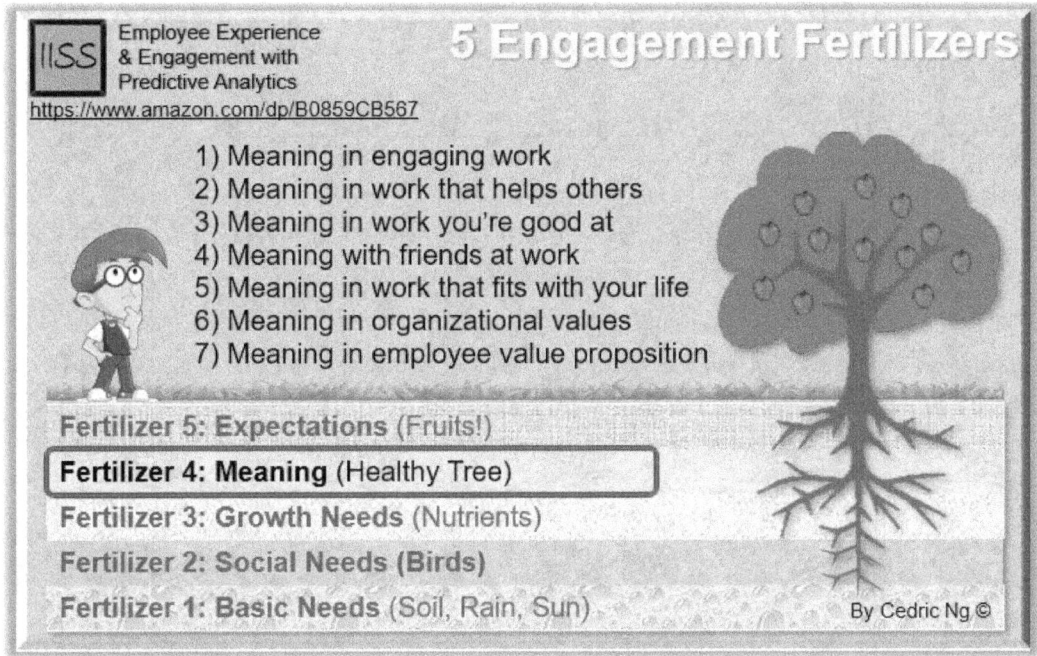

Satisfying personal relationships are a key component for a satisfying life in almost every theory of human needs and well-being (such as Self-Determination Theory and Maslow's hierarchy of needs). [1]

If you hate your colleagues and work for a boss from hell, you are not going to be engaged. Since good relationships are such an important part of having a fulfilling life, it is important to have friends at work. When we think of dream jobs, we usually focus on the company brand and the role. But who you work with is just as important. A bad boss can ruin a dream position, and boring work can be fun if done with a friend. [2]

A study by Walster found that the more similar you are with someone (on almost any dimension — physical appearance, attitudes, personality, interpersonal style, cultural background) the more likely you are to like them. [3]

Reference
(1) Roman Duda (2014) Job satisfaction research https://80000hours.org/articles/job-satisfaction-research/#2-work-that-helps-others (25 February 2020)
(2) Benjamin Todd (2017) We reviewed over 60 studies about what makes for a dream job. Here's what we found https://80000hours.org/career-guide/job-satisfaction/#1-work-thats-engaging (4 October 2019)
(3) Wikipedia (2020) Interpersonal attraction https://en.wikipedia.org/wiki/Interpersonal_attraction#Similarity_attraction_effect (25 February 2020)

3.5.5 Meaning in work that fits with your life

Einstein had his most productive year in 1905, while working as a clerk at a patent office. You don't have to get all the ingredients of a fulfilling life from your job. It's possible to find a job that pays the bills and excel in a side project; or to find a sense of meaning through philanthropy or volunteering; or to build great relationships outside of work. [1]

Reference
(1) Benjamin Todd (2017) We reviewed over 60 studies about what makes for a dream job. Here's what we found https://80000hours.org/career-guide/job-satisfaction/#1-work-thats-engaging *(4 October 2019)*

3.5.6 Meaning in organizational values

People who find meaning in the organizational values, feel that the organization shares and respects their values, and works to achieve goals that are congruent with those values. We find this in religious groups, interest groups, political parties, and charitable organizations, etc.

3.5.6.1 Values that engages employees

For employees to find meaning in their work, they should be able to say, "My values are aligned with the organization's values."

Every business is different, and so are their core values. There are some core values that are commonly used, even though they may be phrased differently. Here are examples of common organizational core values that matters to employees.

(i) Autonomy

People want some autonomy. Autonomy makes people feel respected and valued. When there is autonomy, leadership values your idea and methods, and lets you choose the best way to make things happen. But autonomy is not about leaving people alone. In fact, people don't want to be left alone – they want leadership and boundaries within which they can exercise a reasonable amount of self-direction. It is only in dysfunctional organizations that employees are left to do their work with little input from their managers. That is not autonomy – that is lack of leadership. At the same time, people don't want too much supervision - they want the right mix of checking in and offering help when needed. [1]

Freedom to choose - Employees need autonomy to feel free to choose to choose the work that's meaningful to them, to choose an area to grow, and to be in control of their future.

Home shoring - Airline JetBlue used autonomy to create its customer service experience. JetBlue "home-shored" its telephone service agents, letting them work from home. Home shoring earned JetBlue 9 consecutive J.D. Power and Associates awards for customer satisfaction. According to a Frost & Sullivan report, home shoring draws from a wider range of potential applicants (students, parents, retirees, disabled, etc), and 80% of home shored telephone service agents have degrees. [1]

Shift work - Some organizations allow their employees to determine where, when, how and with whom they will work with. For example, some companies let employees log into a scheduling app to input the shifts which they are available. As long as all shifts are covered, employees can work the shifts and number of hours they have chosen. [1]

Goal Setting - Management is responsible for setting broad goals that impact the organization, business unit, or department. However, employees can be given the autonomy to set their own project milestones.

Tradeoffs - In February 2013, Yahoo's CEO, Marissa Mayer, stunned the company by announcing that employees who had been working full time from home would hence forth be required to come into the office. "To become the absolute best place to work, communication and collaboration will be important, so we need to be working side-by-side." Sometimes an organization must choose to emphasize one driver of engagement over another. In this case, Mayer decided that "connection" was more important to Yahoo's comeback than "autonomy". [1]

(ii) Trust

In 2001, when Douglas Conant took over as Campbell's Soup CEO, it was losing market share, and 62% of its employees were not actively engaged in their jobs. But by 2009, 68% of its employees said they were actively engaged, and Campbell's earnings increased around 4% per year over eight years. How did Conant do it? He made a commitment to his people, embodied in the phrase "Campbell valuing people, people valuing Campbell." He launched programs to get managers communicating with direct reports, and had direct reports evaluate managers. The top criterion that managers were expected to show: the ability to inspire trust. Those who did not measure up were replaced from within. [1]

(iii) Respect

Respecting all employees means respecting their individual human rights and privacy, eliminating all kinds and forms of discrimination, whether based on religion, belief, race, nationality, gender or physical disability. Ensuring a safe and healthy work environment for all employees is also an important part of giving respect to them. Organizations that show respect to its people usually have a lower employee turnover. [2]

References
(1) Tracy Maylett, ED.D. & Paul Warner, PHD (2014) Magic. Five keys to unlock the power of employee engagement (11 January 2020)
(2) Kim Lee and Jenna Arcand (2020) 4 Core Values That Every Organization Should Have
https://www.workitdaily.com/must-have-company-core-values (4 October 2019)

3.5.6.2 Planting values

Provide shareable social media content

People want to know that what is important to them is important to the company. It is the organization's job to communicate its values clearly and consistently. However, often people don't know what the company's values are, let alone how it applies to the work they do. Are your employees proud of being associated with your organizational brand and values?

Companies can create informative content that employees can easily share with their connections on social media. For instance, the company can create an informative blog and request its employees to share daily posts on their social media accounts. This helps potential new hires self-select for fit with the organization's values. Employees will share content and engage customers online if there are rewards for doing so. The incentives can be either monetary or non-monetary. The number of shares, responses to customers, or comments can be used to measure performance for the incentives program. [1]

Align your employee perks with your values

While billiard tables and computer games in office may be the perks that a technology company provides, a Law company or Hospital may offer more opportunities to attend conferences and workshops.

References
(1) Lamont Exeter (2019) Three Ways to Digitize the Workplace to Boost Employee Engagement *https://www.ttec.com/articles/three-ways-digitize-workplace-boost-employee-engagement* (3 October 2019)

3.5.6.3 Planting values activity

Leaders at every level need to identify the values that are most important to the people who report to them. To figure out what drives them, create a list of values and ask your employees (via surveys, interviews or workshop) which are the most important to them as individuals. It is difficult for people to align the organizational values with their personal values if they don't know what their personal values are. The following activity helps people to derive meaning from their work by aligning their values.

1) Look at this list of values

• Accountability • Achievement • Balance (Work/Home) • Commitment • Compassion • Competence • Continuous learning • Cooperation • Courage • Control • Creativity • Customer satisfaction • Dialogue • Driving change • Energy • Engagement	• Enthusiasm • Environmental care • Efficiency • Ethics • Excellence • Fairness • Family • Financial gain • Friendship • Future generations • Goals orientation • Health • Honesty • Humour / fun • Humility • Inclusiveness	• Independence • Integrity • Initiative • Intuition • Loyalty • Involvement • Making a difference • Open communication • Openness • Optimism • Professionalism • Perfection • Performance • Personal fulfillment • Personal growth	• Power • Pride • Prosperity • Quality • Respect • Responsibility • Risk-taking • Spirit • Success • Teamwork • Transparency • Trust • Vision • Wisdom

2) Highlight the 3 Personal Values that guide you in your life

List 3 of your Personal Values:
1.

2.

3.

Why are these Personal Values important to you?
1.

2.

3.

How do people notice that these Personal Values are important to you?
1.

2.

3.

3) Connect your Personal Values to your Organisation's Values

List 3 of your Organisation Values:
1.

2.

3.

How does your Organisational Values connect to your Personal Values?
1.

2.

3.

3.5.7 Meaning in employee value proposition

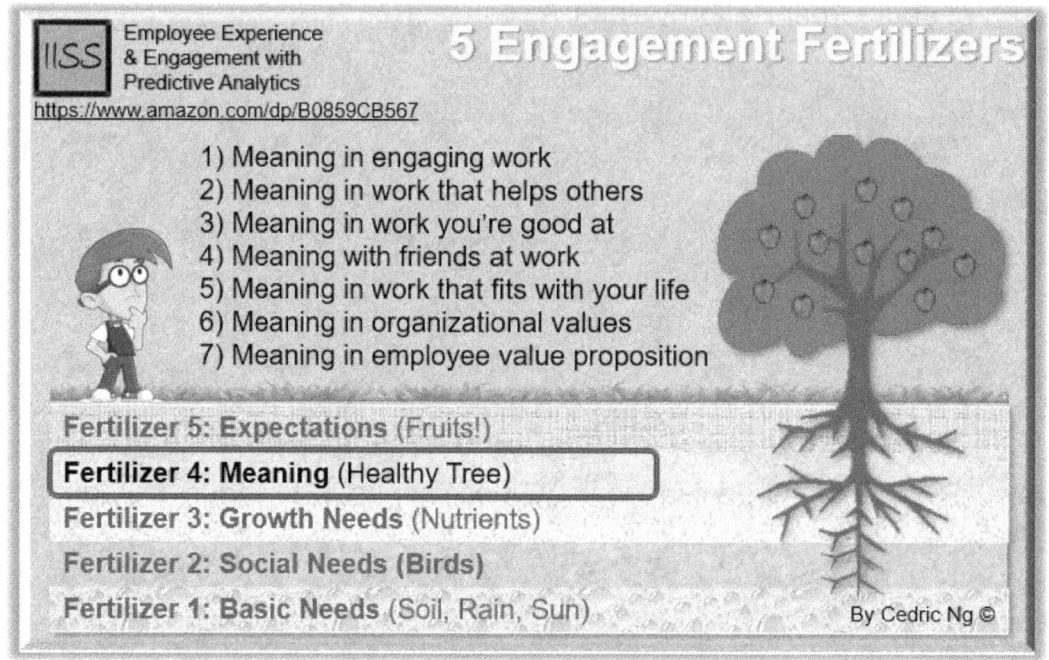

3.5.7.1 What is Employee Value Proposition (EVP)?

The customer value proposition defines the value of a firm's products or services to the consumer. Similarly, the employee value proposition (EVP) is the value (tangible, intangible, and reputational) that an employee gets from an organization in exchange for their work. To discuss the organisation's EVP, start by asking 'Why would people want to work for you?" Is it because of the salary? The benefits? Job responsibilities? Colleagues? Career opportunities? Branding?

3.5.7.2 EVP by industries and jobs

Banking (Goldman Sachs)

Young investment bankers at Goldman Sachs often leave the company after a predictable two-year employment tenure, because of its hard driving culture. In 2015, Goldman Sachs redefined its EVP: [1]

- **Reduced working hours**. It set a clear expectation that interns should leave the office before midnight each evening. This eliminated the perception that junior staff were expected to work the all-nighters for which the position had been famous. It prohibited Analyst from working from 9pm on Fridays to 9am on Sunday.
- **Reduced mundane task**. It reduced some of the work and mundane task by automating them.
- **Provided job rotation**. It launched a 12-month rotation program, so that new employees can gain exposure to other parts of the business.
- Accelerated promotions. The firm accelerated promotions after the second year, along with accompanying increases in salaries.

Call centers

Turnover rates in customer call centers tend to be high because of low pay and repetitive work, and because organizations use the "Basic Needs" EVP of "Come work for us, and we will pay you". A solution for organizations to reduce call center turnover rates is to change their EVP from "Basic Needs" to "Growth Needs". Growth Needs EVPs use phrases such as "career", "tuition reimbursement" and "opportunity to build your resume" in their job advertisements. By the Growth Needs Fertilizer, not only can turnover drop, employee's energy level, and performance will increase.

- **Mentoring opportunities**: Assign mentors to all employees, to provide growth opportunities to both mentor and mentee.
- **Cross-training**: Rather than focusing on a just one product, cross-train employees on a number of products.
- **Problem solving teams**: Give employees challenging work by putting them in problem solving teams.
- **Offer choices:** Give employees a choice as to the products for which they wished to handle calls. Employees will see these as opportunities to work with something that interest them, as well as an opportunity to learn something new.

Ecommerce (Amazon)

According to researchers at the University of Kansas, who surveyed 993 companies across all sectors using data gathered from Glassdor.com, Amazon ranked 63th in overall employee satisfaction. What's strange is that during that same period, Amazon scored much lower in the area of work-life balance. The company never promised a heart-warming experience. But it did promise to add heft to the professional experience section of your resume and put an above-average wage in your pocket. It is not so much whether conditions are good or bad, instead it is more important that expectations are aligned and reasonably being met. Amazon excels in Expectation alignment because it lets new hires know exactly what to expect from their employment and then fulfilled those expectations. That doesn't make the company a good fit for everyone, though: [1]

- Bad fit: You are a family man or woman who wants your job to fit into your need for home time, leisure, and rejuvenation.
- Good fit: You are a millennial starting your career who's okay with a few years of being work-possessed, having a limited person life, and being pushed harder than ever before while building a resume and being well compensated.

Environment (Greenpeace)
Greenpeace's mission is 'to expose global environmental problems, and to force the solutions which are essential to a green and peaceful future". [2]

Health Care (American Red Cross)
American Red Cross's mission is to "To prevent and alleviate human suffering in the face of emergencies by mobilizing the power of volunteers and the generosity of donors." [2]

Health Care (e+CancerCare)
e+CancerCare wanted to create a "build the resume" culture, and published this objective for its employees: "Help our people add a line to their resume every year". [3]

Marketing software automation company (Aprimo)

Aprimo guarantees recent college graduates a promotion within a year through its OnTrack program. As a result, they have increased their retention rate to 85 percent for a five-year period. Gen Y employees will leave your company if they aren't given career opportunities. If they see no path to the top, or no way to gain new experiences, they will leave. [4]

Restaurants (Fast food)

Fast food restaurant employees may be most engaged by free lunches and flexible work schedules that lets them to meet school, family and social obligations.

Restaurants (Upscale restaurants)

Upscale restaurant employees may be most engaged by opportunities for development, growth and satisfied customers.

Restaurants (Fast food - MacDonald's)

McDonald's want their staff to see it not just as a job, but a place where they can find meaning and develop themselves. It promises that every employee will find meaning in working for them. Every crew has the opportunity to be a crew leader and restaurant manager. In fact, 50% of their managers started as a crew. McDonald's is not just a big 'McFamily', it's also a very diverse and inclusive one - they have employees from students to the silver haired generation and even those with disabilities. [5]

Technology (Google)

Google, the king of employee perks, offers way beyond basic benefits to lure top tech talent. Google will feed you, do your laundry, run your errands via Google Shopping Express, etc. In return, you'll have to work long hours, and be the best of the best. Yet, according to payroll consultancy PayScale the median employee tenure at Google is just over one year. [6]

Zoo

Employees can find meaning in seemingly routine and even distasteful work. Zookeepers is probably among the most demanding jobs, as it can be unglamorous (cleaning animal shit) and dangerous (taking care of lions). Although not every aspect of their job was engaging, as a whole their job engaged them deeply. Zookeepers create their own engagement, as they feel their work has a greater purpose of caring for amazing creatures.

References
(1) Tracy Maylett, EdD, and Matthew Wride, JD (2017) The employee experience. How to attract talent, retain top performers, and drive results (11 January 2020)
(2) Alessio Bresciani (2020) 51 Mission Statement Examples from The World's Best Companies https://www.alessiobresciani.com/foresight-strategy/51-mission-statement-examples-from-the-worlds-best-companies/ (11 February 2020)
(3) Tracy Maylett, ED.D. & Paul Warner, PHD (2014) Magic. Five keys to unlock the power of employee engagement (11 January 2020)
(4) Dan Schawbel (2012) 5 Ways to Retain Gen Y Workers https://www.americanexpress.com/en-us/business/trends-and-insights/articles/5-ways-to-retain-gen-y-workers/?fbclid=IwAR0a2431isZHKS5JPeZGmU3GVXsp4pz5qX264XeD4CpdxoV5WfqL7t3Fvl4 (12 March 2020)
(5) Hrmasia (2020) It's one big McFamily at McDonald's https://hrmasia.com/its-one-big-mcfamily-at-mcdonalds/ (11 February 2020)
(6) Leonid Bershidsky (2013) Why Are Google Employees So Disloyal? https://www.bloomberg.com/opinion/articles/2013-07-29/why-are-google-employees-so-disloyal- (11 February 2020)

3.6 Engagement Fertilizer 5: Expectations

To get big and sweet apples, you need to provide the right amount of sunlight and water (alignment of expectations). Different trees need different amount of sunlight and water (different expectations). Too much or too little sunlight and water can kill the apple tree (misalignment of expectations can kill the apple tree).

Similarly, the Organization and Employee have different expectations for Basic Needs, Social Needs, Growth Needs, Meaning, Autonomy, Equity, etc. – Both the Organization and Employee expectations have to be aligned and met for employee experience and engagement to flourish. Employees will soldier on through even the most challenging workloads and environments, if both the Organization and Employee expectations are realistic and aligned.

An employee starts forming expectations before they even become an employee (through contact with your brand). Expectations continue to form during the interview and are constantly being formed throughout the employee lifecycle. Expectations covers compensation, perks, working hours, output, and work environment. A new hire may have 4 main expectations.
1) Basic needs: I don't expect to work on weekends
2) Social needs: The company is fun, easy going place to work.
3) Growth needs: There are opportunities for job rotation and advancement.
4) Meaning: I love the job scope.

3.6.1 Expectancy Theory

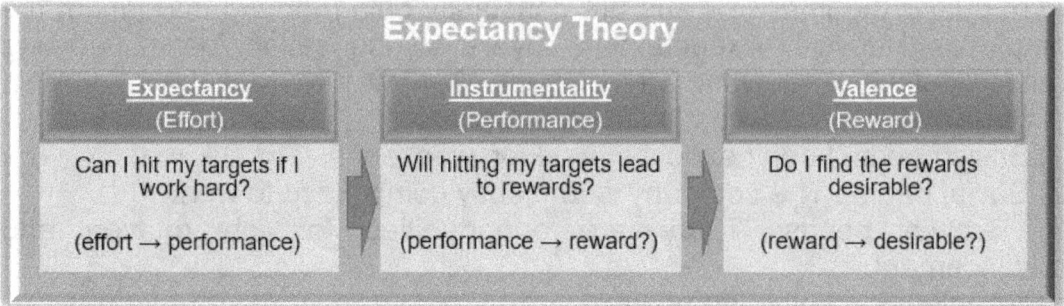

Expectancy theory explains why we are motivated choose one option over others, and is based on three factors (Expectancy, Instrumentality, Valence), each of which must be considered when we look at motivation:[1]

1. **Expectancy** (Effort)

    Expectancy is the belief that one's effort will result in attainment of desired performance goals (effort → performance?). Questions to ask to determine expectancy includes:
    - If I study tonight, will my exam grades improve?
    - If I work harder than everyone else in the factory, will I produce more?
    - If I make more sales calls, will I make any more sales?

2. **Instrumentality** (Performance)

   Instrumentality is the belief that a person will receive a reward if the performance expectation is met (performance → reward?).
   - If I produce more in the factory, will I get a bigger bonus?
   - If I make more sales will I get a bigger commission?
   - If I make more sales will I be recognized by organization as the best sales person?

3. **Valence** (Reward)

   Valence is the value an individual place on the reward (reward → desirable?).
   - Is the extra time I spend making extra sales calls worth the extra commission?
   - Is it important to me that I am the best salesperson?

<u>References</u>
(1) BRIAN FRANCIS REDMOND (2016) *Expectancy Theory* *https://wikispaces.psu.edu/display/PSYCH484/4.+Expectancy+Theory* (23 January 2020)

3.6.2 Equity Theory

The Equity Theory of Motivation, was developed by John Stacey Adams. It is based on the idea that individuals are motivated by fairness. If an individual identifies an inequity between themselves and a peer, they will adjust the work they do to make the situation fair. If an employee found that a who peer does the same job as them is earning more money, they will try to restore equity by changing the actual inputs (do less work) and or outputs (demanding more money), or removing themselves from the perceived equity (leaving the organization). Adam's Equity Theory tells us that the higher an individual's perception of equity (fairness), the more motivated they will be. Conversely, an individual will be demotivated if they perceive unfairness. [1]

Initially, a new hire may not expect to receive any bonus in their first year, as per company policy. But if they see other new hires, or lazy colleagues get bonuses for "special achievements". The employee will also expect special bonus for their work, especially when they evaluate their contributions against others. Their expectations have changed. The new hire now expects a bonus – it doesn't matter whether this expectation is logical.

Inputs and Outputs

To understand Adam's Equity Theory, we need to define inputs and outputs. [1]

(i) Inputs: Inputs are things that an individual does in order to receive an output. They are the contribution the individual makes to the organization. Examples of inputs include:
- The number of hours worked (effort).
- Personal sacrifices made.
- The loyalty the person demonstrated his manager or the organization.
- The flexibility shown by the person, by accepting assignments with tight deadlines.

(ii) Outputs: Outputs are the result an individual receives as a result of their inputs. Some of these benefits will be tangible, such as salary, but others will be intangible, such as recognition. Examples of outputs include:
- Salary
- Bonus
- Company car
- Stock options
- Recognition
- Promotion

Understanding Equity

Equity is an individual's outputs divided by that same person's inputs. Individuals don't just see equity in isolation, instead they compare themselves to others. If they perceive an inequity then they will adjust their inputs to restore balance, as show in the following equity theory equation.

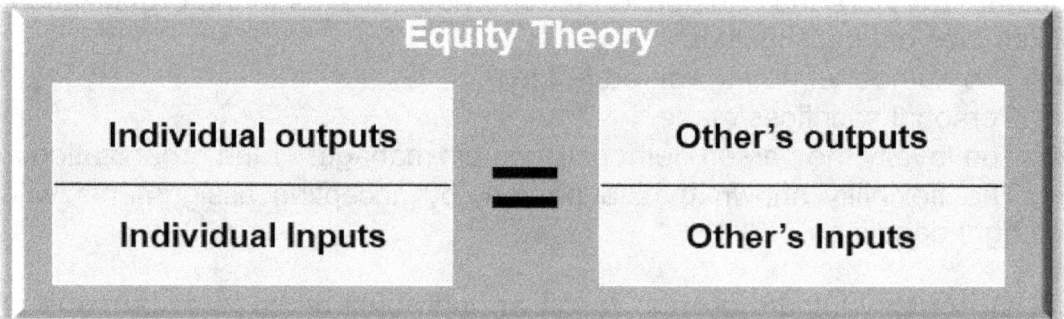

References
(1) expertprogrammanagement (2017) Equity Theory – Keeping Employees Motivated https://expertprogrammanagement.com/2017/06/equity-theory/ (11 February 2020)

3.6.3 Expectation Gap

DecisionWise found that engaged teams had clear expectations. In fact, the need for clear expectations out weighted factors like compensation, working conditions, perks, and training. The only other area that came close to the need for "clear expectations" was "recognition". Recognition and expectations go hand in hand - it's difficult to recognize and reward good performance if you don't know the targets are. [1]

Expectation gap is the difference between what employers say and what employees believe. There will always be a gap between what was promised and what people think they were promised. The manager's job is to be aware of the gap and manage it. Wise leaders, "under promise and over deliver" – because an employee who gets a surprise salary bonus will be very happy if he expected to receive nothing.

Some people work long hours. Some people work with dangerous prisoners. Some people work in unpleasant and unsafe locations. Some people spend months on oil rigs in the middle of the ocean. Firefighters willingly put their lives on the line every day for strangers, proudly and with passion. - Yet these people love their job and are highly engaged because they got what they expected. Engagement depends on whether expectations are aligned and met.

Many organizations operate under the misguided assumption that money is an employee's main motivator, and spend millions on staff perks. They are wasting their money or "digging in the wrong place", if those perks weren't what the employee signed up for. Expectation gaps are often the reason for employee disengagement. People who leave usually complained about unmet expectations, and use phrases like:
- I thought I would have been promoted.
- I didn't know that I have to work till midnight every day.
- I didn't know that this is what I will be doing every day.

<u>References</u>
(1) Tracy Maylett, EdD, and Matthew Wride, JD (2017) The employee experience. How to attract talent, retain top performers, and drive results (11 January 2020)

3.6.4 Roles of management & employees in aligning expectations

Both management and employees have a role in aligning expectations

Expectations of Management

The organization is responsible for building an environment (salary, benefits, and overall working environment) where engagement can grow – tilling the soil so that engagement can grow. As a manager you have more influence than what you may realize over the engagement of the people you manage. The more engaged you are as a manager, the more engaged your people will be. Being engaged as a manager is different from being engaged as an employee. Managers have the biggest impact on engagement, because they shape their work experience every day. The engaged manager tills the soil for his people, creating the same conditions that got him engaged in work. He then gets his people engaged in the ways that are unique to them, based on what he knows about their interests and needs.

Expectations of Employee

While management provides the essential ingredients, it is ultimately up to the employee to choose to be engaged. If the organization is a plant, then each employee is like a single cell. Change may appear on an entire plant, but change begins at in a single cell. An employee's greatest impact on engagement in their organization is how engaged they are personally. Although the organizational and managers provide the opportunities for employees to find social needs, growth and meaning at work – it is up to the employee to want to be engaged.

Engagement is a choice

We have seen people mundane, repetitive even unpleasant jobs, and watched them turn it meaningful because they chose to be engaged in what they were doing. It's employee's choice to be engaged or disengaged. Employees can be engaged even if their employer doesn't create an environment that's conducive to engagement.

Employees create their own meaning

Meaning is highly individual. Managers can till the soil for meaning to grow by having a clear organizational mission, sharing the corporate story, and explain to everyone "this is why we do what we do". However, employees themselves need to connect the organizational meaning to their personal meaning. But nobody can tell you what's meaningful to you.

3.6.5 How to align expectations?

There are several ways to align the expectations of management and employees:
- Set clear mutually agreed objectives
- Acknowledge employee's struggles
- Give Feedback and Ask for Feedback
- Conduct stay interviews
- Practice humility

Set clear mutually agreed objectives

It is important to communicate and align business, team, and individual objectives, so that everyone pulls in the same direction.

Employees need to know what is happening within their organization to be able to align their individual work goals. Do they know what are the company goals and how the company is doing? Do they know what are the challenges and opportunities that the company is facing? Employees who are kept in the loop and can see how their work contributes to the overall success of the organization will be more motivated to achieve their individual work goals. Foster greater transparency at work by updating employees on both formal and informal matters and, share the company financials with employees. Being able to clearly see the role they play in overall company success helps create a better sense of ownership. Engagement comes from thinking and acting like owners. [1]

Give "inside" information - To get your staff more involved and committed, keep them up to date with "inside" information, such as the direction of the company and the challenges that the Leadership Team is facing. Let your people know that it's "inside" information. Trusting your employees can handle it raises engagement. People want to be in the know, they want to be in the circle of trust.

Acknowledge employee's struggles

Most employees hide their struggles for fear of looking incompetent. Acknowledging their struggles makes your gratitude more credible, and makes it safer for employees to be honest with you in the future when facing difficulties. [2]

Nothing affirms an employee's great work more than a leader saying, "That was amazing. Tell me how you did it?" By honoring the story behind the work, help them feel that they, and their work, really matter. [3]

Conduct stay interviews

Managers can schedule stay interviews with employees on their one-year mark so they know what is important to them and can influence their decision to stay. Good questions to ask include: [4]

1) "What do you like about your job?"
- helps a manager know what parts of the job employees like and want to experience more of.

2) "Describe a good day of work you had recently?"
- ask this question to replicate the good experience.

3) "Do you feel your skills are being utilized to the fullest?"
- discovering that the employee has skills the company never knew about is a win-win.

4) "Do you feel you get properly recognized for doing good work?"
- recognition for accomplishments is linked to higher employee retention.

5) "Do you feel like you are treated with respect?"
- lack of respect is a contributor to employees leaving.

6) What do you want to be doing that you aren't currently doing?
- helps a manager know how to redesign their employee's job scope to motivate them.

Give Feedback and Ask for Feedback

A manager needs to provide regular feedback to their employees, and ask their employees for feedback.

According to a Gallup study, organizations who conduct regular employee feedback have turnover rates that are 14.9% lower than for employees who receive no feedback. [5]

Contributing to the improvement of a company is an often-overlooked engagement strategy. Employees often have good ideas for improving products, services, systems, and processes. But it can be frustrating for them to have a solution to a problem but not given any feedback channels to voice it. Suggestion boxes or emails are examples of feedback channels.

Practice humility

Source: https://corporate-rebels.com/iceberg-of-ignorance/

Leaders who show humility by mixing with the front-line gain more status and influence than their peers who prefer to stay in their offices. It is difficult for the leadership team to solve organization problems, especially if they are only aware of the 'tip of the iceberg'. For leaders at all levels, humility is the key to melting the iceberg. Harvard Business School Professor Francesca Gino found managers with the least levels of respect are also those known for shutting themselves in their offices. When Sidney Yoshida produced his study called 'The Iceberg of Ignorance', he found that although 100% of front-line problems were known to the front-line employees, only 74% were known to team leaders, 9% to middle management and just 4% to top management! [6]

References
(1) Andre Lavoie (2015) The 5 Leadership Behaviors You Need to Boost Employee Engagement https://www.entrepreneur.com/article/247099 (4 October 2019)
(2) Lamont Exeter (2019) Three Ways to Digitize the Workplace to Boost Employee Engagement https://www.ttec.com/articles/three-ways-digitize-workplace-boost-employee-engagement (3 October 2019)
(3) Ron Carucci, HBR (2019) A recipe for disengagement on your team. HBR Guide to Motivating People. United States of America. Harvard Business School Publishing Corporation.
(4) Marcel Schwantes (2017) Want Your Best Employees to Never Leave You? Ask Them 5 Simple Questions https://www.inc.com/marcel-schwantes/want-your-best-employees-to-never-leave-you-ask-them-5-simple-questions.html?cid=sf01002&sr_share=facebook&fbclid=IwAR0azTnTInUoRszPFRw6P_BEqpuF-GMkBpSbIp2xHySSU9vXKJ2TqzGfX94 (4 October 2019)
(5) Jim Asplund and Nikki Blacksmith (2011) The Secret of Higher Performance https://news.gallup.com/businessjournal/147383/secret-higher-performance.aspx (3 October 2019)
(6) Joost (2011) How Real Leaders Melt The Iceberg of Ignorance With Humility https://corporate-rebels.com/iceberg-of-ignorance/ (6 January 2020)

4.0 Engagement Bag 3: Sentiment Gathering

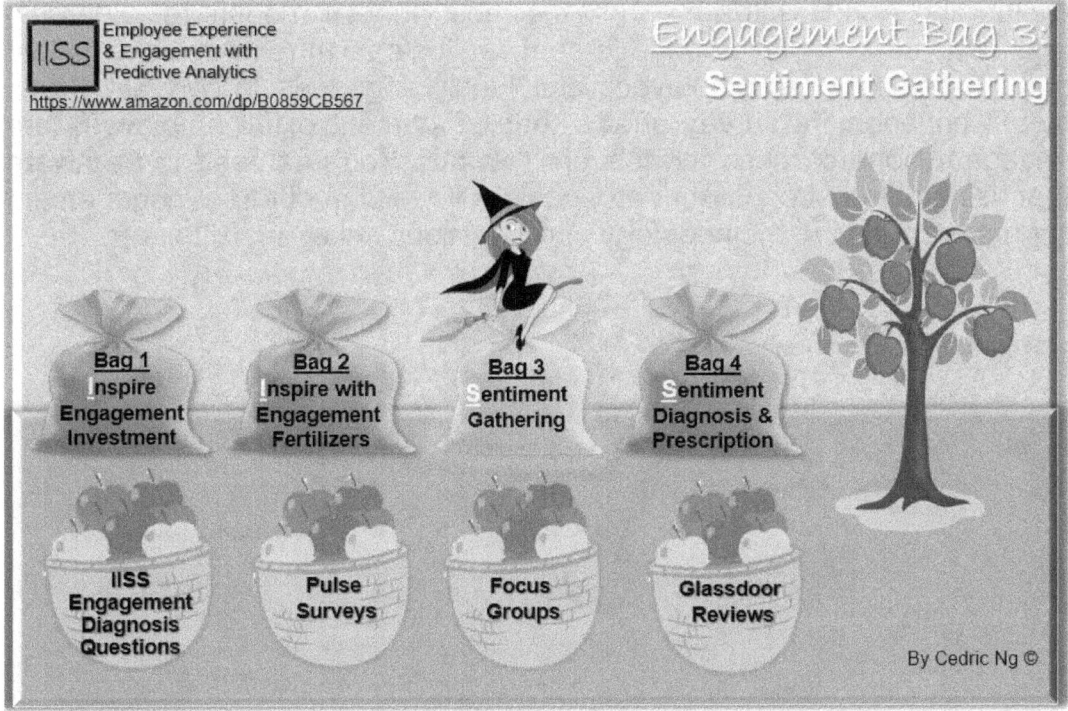

In IISS, "**Sentiment Gathering**" is the third engagement bag because you can start evaluating the effectiveness of the engagement programs after you've launched it. There are various ways to gather employee sentiment. This section covers these approaches to gather employee sentiment:
- IISS Engagement Diagnosis Questions
- Pulse surveys
- Focus groups
- Glassdoor reviews

Different engagement models have different engagement scores

Gallup and Aon Hewitt are two of the most popular models for gathering and measuring employee engagement. Gallup's 2014 research shows that only 13 percent of all employees are "highly engaged". In contrast, Aon's global engagement survey shows that 65 percent of all employees are engaged. None of these models are "wrong". You just need to be aware that using different engagement models will result in different engagement scores because the methodology and questions asked are different.

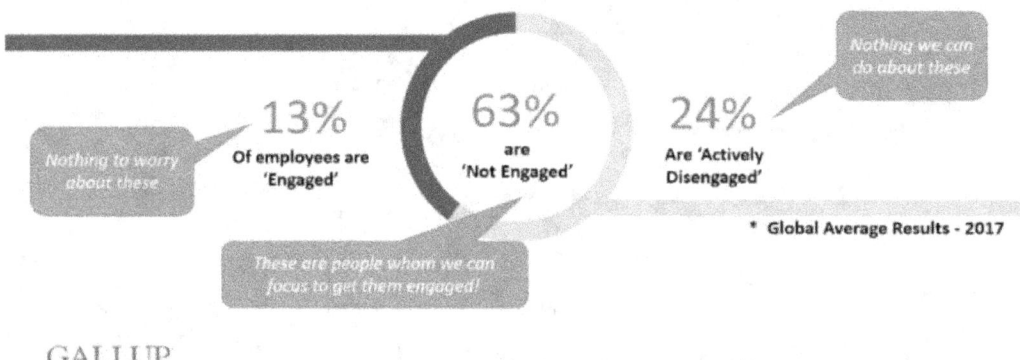

Source: https://acuvate.com/blog/top-9-actionable-employee-engagement-ideas-and-activities/

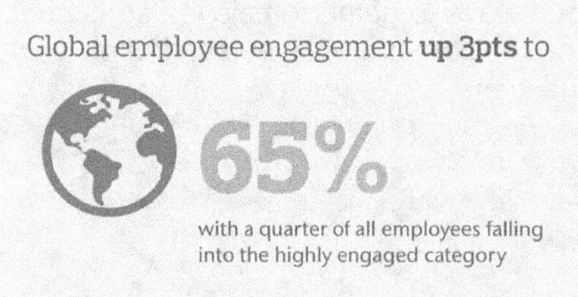

Source: https://www.aonhewitt.co.nz/Home/Resources/Reports-and-research/2016-Trends-in-Global-Employee-Engagement

Different countries have different "average" engagement scores

Different countries have different "average" engagement scores. Employees in Malaysia and Singaporean are the least engaged among major Asian markets. Engagement scores for India are 69%, followed by China (67%), Thailand (65%), Philippines (65%), Indonesia (61%), and Malaysia (59%). Aon's analysis found regional variations in engagement are driven by regional and country-specific economic, political and cultural differences.

Source: https://apac.aonhewitt.com/home/about/media-room/press-releases/may-2017

4.1 Organisational engagement survey

Flaws in survey planning and design leads to bad results. This section covers how to plan and design engagement surveys, so that the right issues can be identified and fixed.

Designing your own engagement survey can be tedious, but if you are reviewing previously developed questions, the process will be much easier. When you create your own engagement survey, you'll want to measure engagement itself (outcome measure), as well as the outside influences (engagement drivers) that may influence engagement in your organization. As you review various engagement drivers, you may decide that not all of them are relevant to your organizational needs.

1. **Determine who will review and approve** the questions. For global engagement surveys, it is good practice to get the regions, countries, and business units to review the questions, so as to get their buy-in. Try not to have too many people approving it, as the sign-off process will become lengthy.

2. **Involve key leaders** to prioritize the questions.

3. **Review the engagement drivers' categories**, and remove those that are not relevant to your organization, and add any new ones if necessary. Engagement drivers (basic needs, social needs, development needs, etc) are factors that influence engagement (outcome) in your organization.

4. **Review the engagement questions**, and adjust the questions to fit your organisational needs. How you structure your survey questions is important.
   - **Social desirability bias**. If you ask people to respond to questions like "I feel overworked" or "I feel burned out" – they are likely to say "no" as it focuses on them rather than the organization or work. It's better to ask people to respond to a question like "Generally, I believe my workload is reasonable for my role", as it focuses on the organization or work, rather than themselves. [1]

- **Acquiescence bias**. People have a tendency to say they "agree" as a default response to survey questions, particularly when their knowledge is limited or none of the available answers fit. [1]
- **Double-barrel questions**. Double-barrel questions are statements with two unrelated components, such as "I am motivated to perform my best work and we are good at holding people accountable." Instead, people should be asked two different questions: "I am motivated to perform my best work in this organization" and "We hold ourselves and our team members accountable for results". [1]
- **Ambiguity**. Double negatives is an example of ambiguity that leave people unsure what the question even is ("I don't feel that my company fails to provide adequate resources to enable me to do my job"). [1]

5. **Include definitions for the engagement survey**. For example, include a definition of your manager, your team, senior leadership, this organization, etc. to ensure people use the same frame of reference when responding.

6. **Send your questionnaire for review and approval.**

<u>References</u>
(1) Jennifer Cullen, HBR (2019) Where engagement surveys go wrong. HBR Guide to Motivating People. United States of America. Harvard Business School Publishing Corporation.

4.1.1 Gallup survey questions

Why ask your employees hundreds of questions to measure their employee engagement if you can get the most accurate results by asking only 12? Gallup researchers spent decades testing hundreds of questions, because their wording and order mean everything when it comes to accurately measuring engagement. Gallup's Q^{12} survey is an effective measure of employee engagement and its impact on the outcomes that matter most to the business. Gallup's Q^{12} survey and management strategies tie directly to increased Productivity, Profitability, and Employee retention. Gallup's Q^{12} survey is available in over 30 languages, and scores are on a 1 to 5 scale. [1]

Gallup's survey measures actionable items that are predictive of attitudinal outcomes such as satisfaction, loyalty, and pride. Similar to Maslow's hierarchy of needs, Gallup's survey uses a four-level employee engagement hierarchy. Employees are asked 12 questions designed to elicit their assessment of each of four levels of engagement. [2]

Source: http://www.educate78.org/why-teacher-engagement-matters-in-oakland/

Gallup's research demonstrates that the first six elements of engagement (Questions 1-6 on the Gallup Q12 survey), including basic needs and management support, are foundational to satisfaction, performance, and retention. The research warns that leaders must ensure that a solid foundation is embedded in the workplace before concentrating on the upper levels of the engagement hierarchy. (2)

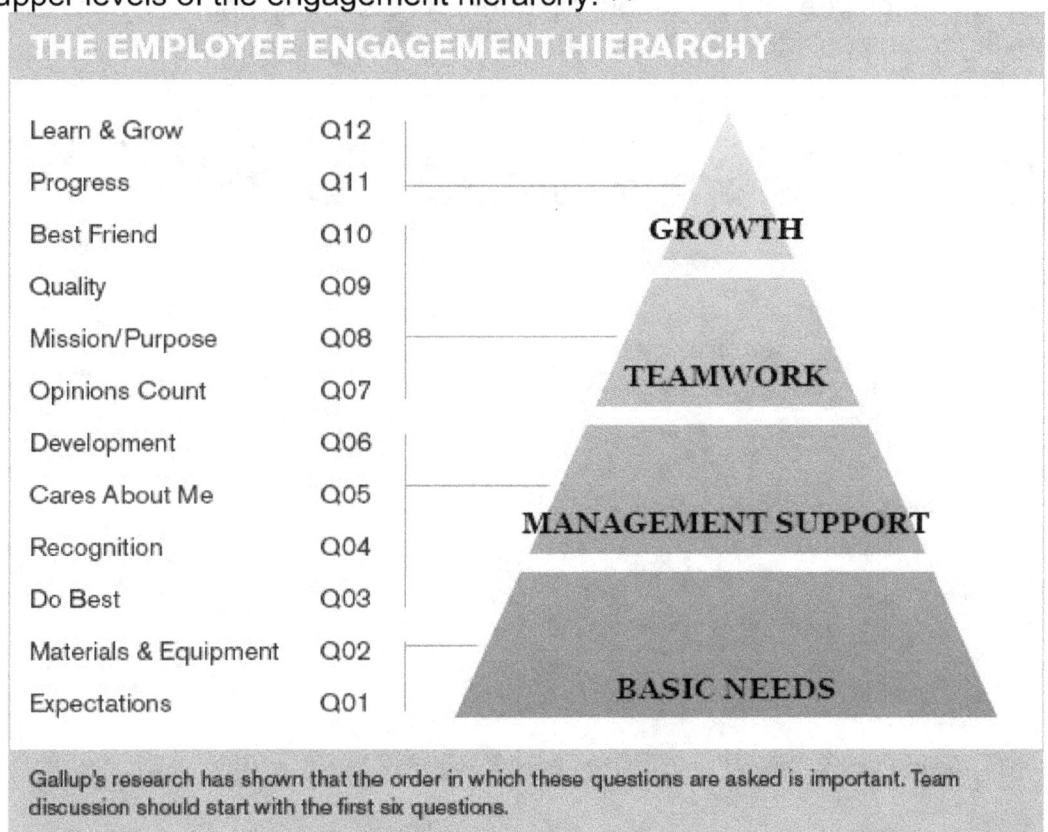

Source: https://sa.berkeley.edu/sites/default/files/images/Gallup%20Pyramid.jpg

References
(1) Gallup (2019) The right questions https://q12.gallup.com/Public/en-us/Features (2 October 2019)
(2) TeachOakland Advisory Group (2016) Why Teacher Engagement Matters in Oakland http://www.educate78.org/why-teacher-engagement-matters-in-oakland/ (2 October 2019)

4.1.2 Aon Hewitt survey questions

Aon Hewitt' Engagement Model, has 6 engagement drivers categories (Brand, Leadership, Performance, The Work, The Basics, Company Practices), and 3 engagement outcomes (Say, Stay, Work). Aon believes that an employee must exhibit all three facets of saying, staying, and striving to be considered "engaged.

Source: https://www.communication-director.com/sites/default/files/aon_hewitt.jpg

Engagement	Content	Items in Aon Hewitt Operational Definition
	Speak positively about the organization to coworkers, potential employees, and customers	• I would not hesitate to recommend this organization to a friend seeking employment • Given the opportunity, I tell others great things about working here
	Have an intense sense of belonging and desire to be part of the organization	• It would take a lot to get me to leave this organization • I rarely think about leaving this organization to work somewhere else
	Are motivated and exert effort toward success in their job and for the company	• This organization inspires me to do my best work every day • This organization motivates me to contribute more than is normally required to complete my work

Source: *https://www.aonhewitt.co.nz/getattachment/77046028-9992-4d77-868a-32fbf622fec6/file.aspx*

AON's Engagement Questions

Engagement	It would take a lot to get me to leave…
	I rarely think about leaving…
	I wouldn't hesitate to recommend this organization to a friend…
	I tell others great things about working here…
	This organization inspires my best work…
	This organization motivates me …
Manager	My manager sets clear goals…
	My manager provides feedback…
	My manager recognizes efforts…
Collaboration	Coworkers work together…
	Coworkers share best practices…
	Good cooperation between departments…
Senior Leadership	Senior leadership communication…
	Senior leadership is accessible…
	Senior leadership provides clear direction…
Rewards	Performance impacts pay…
	Paid fairly…
	I receive recognition…
Enablement	Required staffing levels…
	Tools and resources support productivity…
	Work processes support productivity…
Learning and Advancement	Support for learning and development…
	Opportunity to gain new skills…
	Career opportunities…

Source: https://www.aonhewitt.co.nz/getattachment/77046028-9992-4d77-868a-32fbf622fec6/file.aspx

4.1.3 Officevibe survey questions

Officevibe found that there are 10 metrics (Recognition, Feedback, Happiness, Personal growth, Satisfaction, Wellness, Ambassadorship, Relationship with managers, Relations with colleagues, Company alignment) that you should measure.

THE 10 METRICS OF EMPLOYEE ENGAGEMENT

For an employee to be engaged, there needs to be a few core things that are taken care of.

Employees need to feel like they're respected, that they're part of the team, and that their ideas matter. Once that's taken care of, they'll be more likely to go above and beyond for their company and provide amazing service, come up with innovative ideas, and help the company grow.

There are 10 essential things that companies need to keep in mind when trying to improve employee engagement.

 1. RECOGNITION
 6. WELLNESS
 2. FEEDBACK
 7. AMBASSADORSHIP
 3. HAPPINESS
 8. RELATIONSHIP WITH MANAGERS
 4. PERSONAL GROWTH
 9. RELATIONSHIP WITH COLLEAGUES
 5. SATISFACTION
 10. COMPANY ALIGNMENT

Source: https://www.officevibe.com/blog/10-pillars-employee-engagement-infographic

1) Recognition

Employees need frequent recognition. Recognition is simply the acknowledgment of a job well done. Employees need frequent recognition. They need to know that their work is good and that you're appreciative with their contribution to the team. According to Gallup, you should be praising your employees at least once every week. Question to ask employees: Do you feel like you get enough recognition for your work? [1]

- Deloitte found that: "Organizations with recognition programs which are highly effective at enabling employee engagement had 31% lower voluntary turnover than organizations with ineffective recognition programs". [1]
- Officevibe found that employees value praise more than gifts - 83% of employees think it's better to give someone praise than a gift. [1]
- According to Gallup research, Variation in response to the Gallup's engagement survey question 12, "In the last seven days, I have received recognition or praise for doing good work" is responsible for a 10% to 20% difference in revenue and productivity. Employees who report that they're not adequately recognized at work are three times more likely to say they'll quit in the next year. It's important to note that Gallup's Q12 assessment that ask if the employee has received praise in the last seven days, may seem like a short time frame, but it takes repeated exposure to build the reward/repeat loop. An annual pat on the back doesn't feed emotional engagement. [2]

2) Feedback

Feedback is critical for employees to grow. Feedback is about changing behaviour, and the closer you give feedback on the behaviour you want to be changed, the more likely it is to change. Monthly one-on-ones are great, but don't forget to have informal sessions. It is important to give quality feedback frequently. Quality feedback are specific and actionable. Officevibe found that 83% of employees appreciate receiving feedback, regardless if it's positive or negative. Question to ask employees: On a scale from 0-10, rate the quality of the feedback you receive. [1]

3) Happiness

Happiness is an important metric to be tracking, but where a lot of organizations go wrong is that they only measure happiness. A study by economists at the University of Warwick found that happiness made people around 12% more productive. You'll likely find out that what employees really want to make them happy are free things, things like more work-life balance, more autonomy, exciting projects to work on, etc. Question to ask employees: On a scale from 0-10, how happy are you at work this week? [1]

4) Personal growth

This is one of the most important metrics. Employees want growth in terms of more money, more responsibility, more credit, more autonomy, etc. Question to ask employees: Do you feel like you have enough opportunities for growth? Follow up this question with a "why". [1]

5) Satisfaction

There are two sides to employee satisfaction. The first, is compensation (salary + benefits), and the second, is their overall work environment. Employees need to feel like they're being fairly compensated for the work they do, otherwise they'll be disengaged. Managers should be frequently checking in with employees to see if they have what they need and if there's any confusion they could clear up. Question to ask employees: Is your work environment distracting? [1]

6) Wellness

Wellness, both physical and mental, are important things to look at. Most employees around the world are stressed, tired, and overworked. Question to ask employees: On a scale from 0-10, what's your stress level at work this week? [1]

7) Ambassadorship

The metric Ambassadorship metric is measuring loyalty and pride. Ambassadors of the brand is the ultimate goal in engagement, when someone asks one of your employees where they work and their eyes light up as they passionately describe their job. This metric is modelled after the Net Promoter score. Question to ask employees: On a scale from 0-10, how likely are you to recommend our organization as a place to work? [1]

8) Relationship with managers

People join organizations but they leave managers. The employee-manager relationship is so important because of the effect a manager can have on an employee. Managers should communicate frequently and politely with their team. Question to ask employees: Is there anything that you would change to improve the relationship with your manager? [1]

9) Relations with colleagues

This is an important metric because teams need to be able to work well together to be productive. It's the leader job to make sure everyone feels connected in the team. Question to ask employees: On a scale from 0-10, how would you rate your relationship with your peers? [1]

10) Company alignment

When an employee's personal values align with the organization's values, you have a true culture fit. That's when you'll have a much easier time motivating and engaging employees because they believe what you believe. One way to help align employees is by helping them understand why you do what you do. Let them see who your customers are, who your competitors are, and what you're doing to differentiate. Question to ask employees: How well do your personal values align with our core values? [1]

References
(1) Officevibe (2017) The 10 Employee Engagement Metrics That Matter https://www.officevibe.com/blog/10-pillars-employee-engagement-infographic (1 October 2019)
(2) Jennifer Robison (2006) In Praise of Praising Your Employees https://www.gallup.com/workplace/236951/praise-praising-employees.aspx (2 October 2019)

4.1.4 IISS Engagement Diagnosis Questions

In the IISS model, there are suggested "engagement diagnosis questions" for each of the "5 Engagement Fertilizers". Engagement diagnosis questions are survey questions to measure employee engagement.

IISS Model's Engagement Diagnosis Questions & Prescriptions	
Engagement Fertilizers	IISS Engagement Diagnosis Questions
Fertilizer 1) Basic Needs	Q1) I believe I am paid fairly
	Q2) I am satisfied with my company's employee benefits
	Q3) I am satisfied with the work life balance here
	Q4) The resources & processes support productivity
	Q5) The physical working environment is appealing & conducive for work
Fertilizer 2) Social Needs	Q6) I work well with my manager
	Q7) I work well with my colleagues
	Q8) I feel recognized for my work
Fertilizer 3) Growth Needs	Q9) I have opportunities to learn and grow
Fertilizer 4) Meaning	Q10) My job is meaningful
Fertilizer 5) Expectations	Q11) My job expectations are aligned with the organization
	Q12) I will recommend this organization to my friends
	Q13) My manager sets clear goals aligned with company

https://www.amazon.com/dp/B0859CB567

By Cedric Ng ©

4.2 Pulse surveys

The employee pulse survey is a tool often used by companies to get a sense of the "health" of the company by evaluating employee satisfaction, productivity, and overall attitude on periodic basis. A pulse survey is different than an employee survey in that a Pulse survey is brief, more specific in goal and is conducted on pre-defined employee groups.

Some organizations such as General Electric (GE) run constant surveys to random employees. E.g. They might do a survey for 1,000 employees one month, and then another 1,000 the next month, and so on. The idea is to get employee feedback constantly from a representative sample size of the organization. For organizations like GE which has more than 300,000 employees, this approach makes more sense than trying to do one annual survey. [1]

Cisco got rid of its annual employee engagement survey and shifted its approach to focus on engagement at team level. Today, any team leader at Cisco can run an eight-question pulse survey anytime he or she wants to get an idea of what's going on in the team. These responses are then analysed and reported back within six days instead of six months. [1]

Reference
(1) Jacob Morgan (2017). The employee experience advantage. Wiley.

4.3 Focus groups

After an organization has conducted an employee engagement survey, employee focus groups may be required to unearth the root causes of issues for action planning. An employee focus group is a group of employees who assemble to participate in a guided discussion about a particular topic. Each individual department, can conduct their own employee focus groups and create their own employee engagement action plans. Each group should have around 10 employees so that everyone has the opportunity to speak. A moderator will lead the group to discuss survey results, and probe the group for sources of issues and solutions for improving employee engagement. [1]

Reference
(1) Hilary Wright (2017) Employee Focus Groups: Your Superpower for Improving Employee Engagement https://www.quantumworkplace.com/future-of-work/employee-focus-groups-your-superpower-improving-employee-engagement/ (3 October 2019)

4.4 Glassdoor reviews

Employee's sentiment about a company can be assessed from the company's rating at Glassdoor, where employees can rate what it's really like to work inside, using a 5-point scale ranging from "Very Dissatisfied" to "Very Satisfied".

Glassdoor (https://www.glassdoor.sg/) is a website whereby you can search the ratings and reviews of over 600,000 companies worldwide. At Glassdoor's website, you can find out what it's really like to work inside any company, from people who've actually worked there. Company ratings on Glassdoor are determined by recent employee feedback. [1]

Company ratings are based on a 5-point scale:

- 0.00 - 1.50 Employees are "Very Dissatisfied"
- 1.51 - 2.50 Employees are "Dissatisfied"
- 2.51 - 3.50 Employees say it's "OK"
- 3.51 - 4.00 Employees are "Satisfied"
- 4.01 - 5.00 Employees are "Very Satisfied"

Example of a Glassdoor Company Ratings and Reviews,

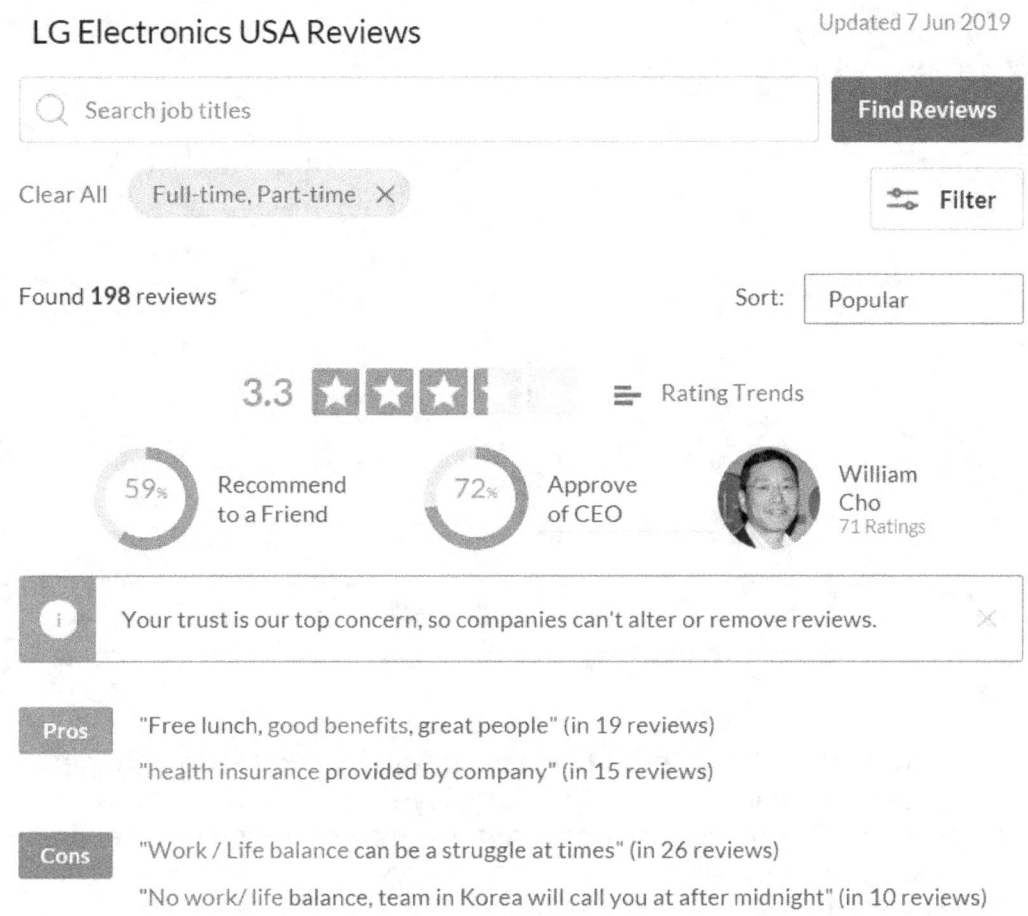

Source: https://www.glassdoor.sg/Reviews/LG-Electronics-USA-Reviews-E276566.htm

Reference
(1) Glassdoor (2019) Ratings on Glassdoor. https://help.glassdoor.com/article/Ratings-on-Glassdoor/en_US/ (15 June 2019)

5.0 Engagement Bag 4: Sentiment Diagnosis & Prescription

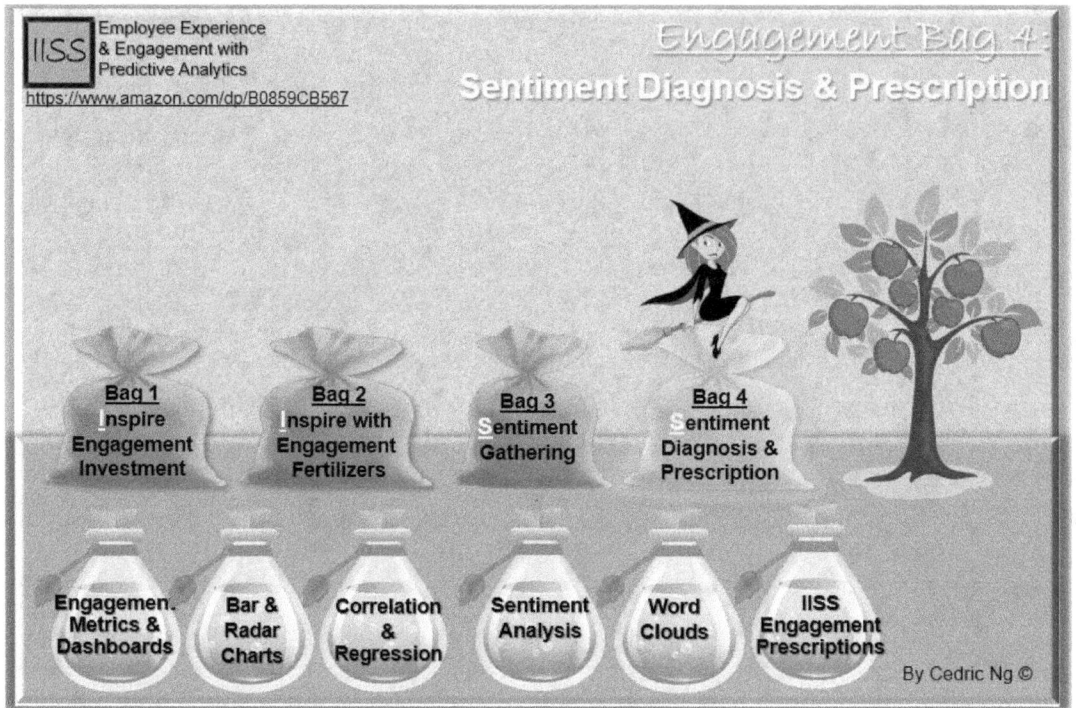

In IISS, **"Sentiment Diagnosis & Prescription"** is the fourth engagement bag because we can only analyse employee sentiment after we have gathered them. There are various ways to analyse employee sentiment. This section covers these approaches to analyse employee sentiment:
- Engagement Metrics & Dashboards
- Bar & Radar Charts
- Correlation & Regression
- Sentiment Analysis
- Word Clouds
- IISS Engagement Prescriptions

5.1 Steps to analyze engagement surveys

Survey responses is a useful litmus test for general sentiment. It is important for organisations to listen to their employees, but listening isn't enough. These are the steps on how to analyse survey results and improve employee engagement.

1. Analyse the Results

Before you start sharing your results, you need to make sense of the data you've been given. These are some ways to analyse the results:

- **Lowest scoring questions.** analysing lowest scoring questions is where you can move the needle on engagement most.

- **Questions with the largest year-over-year change.** if you have multiple years of survey data, analyse areas where engagement has changed significantly. It might be that external factors that affected the results, if there isn't major changes in your organisation.

- **Questions where a team scored below the average.** break the data into smaller groups by region, country, or departments so you can pinpoint the areas to address. This can help you to pinpoint where problems and solutions lie. If the engineering department has low engagement, but the Purchasing department's engagement is very high, are there things that the Purchasing department is doing differently that could transfer? Do you need to hire more engineers?

- **Make hypotheses.** Before you share the survey results, brainstorm potential solutions to those problems. Employees are going to ask questions about what you're planning and what your next steps are.

2. Share the results

After analysing the results, you're ready to share the results. To finally present your work to the team:

- **Thank them for their contributions**. This should be your opening, especially if your survey was voluntary.

- **Let employees know you've identified areas of concern**. This shows that you do pay attention to employee feedback.

- **Outline your basic plans and next steps**. This makes it clear that you aren't just listening to concerns, you are actually addressing them.

- **Ask for feedback**. This ensures that employees are involved and feel that they have a role in shaping their own engagement. Employees can give you insights that can better your plans. Employee feedback can also help you to see if the areas that you have identified as problematic resonates with them.

3. Discuss results and Ideas for improvement
Discuss the organizational and team results with the managers to get their input on what could be causing the problems. Get them to identify two or three areas to focus, and identify possible solutions.

4. Implement the solutions
Taking action can be overwhelming, if it involves major changes. To make the process more manageable, break up larger initiatives into smaller steps for multiple individuals, ensuring the stakeholders are involved and accountable.

5. Evaluate
Hold the stakeholders accountable by routinely updating progress publicly. Evaluate if the team's agreed-upon actions don't seem to be working, and discuss alternative actions.

5.2 Engagement Metrics & Dashboards

5.2.1 Engagement Metrics

Principles of metrics application

There are numerous types of metrics that can be used to measure impact of HR strategies. Huselid, Becker and Beatty (2005) introduced five key principles of metrics application to measure HR strategies:
- **Principle 1**: The metrics must help answer the organization's strategic questions
- **Principle 2**: Determine metrics that are valid to strategies, then sieve out those which are also feasible for measuring.
- **Principle 3**: Metrics chosen should be holistically related to one another, so that they can reveal larger pictures. Do not drill down to detailed level such that relationships and overall pictures are lost.
- **Principle 4**: Select metrics that involves responsibilities of HR and all other functions
- **Principle 5**: Avoid numerous metrics and focus only on the essential few that will help answer key strategic questions.

When employees are not engaged, you are likely to see the following:
- Increase in absenteeism
- Increase in grievances
- Increase in employee turnover rate
- Decrease in productivity

5.2.1.1 Increase in absenteeism

Absenteeism Rate
Formula: Sick Days / Headcount
Metric Description: The number of work days missed due to illness per Headcount. An increase in absenteeism may also indicate a disengaged workforce and can be considered a leading indicator of future employee turnover.

5.2.1.2 Increase in grievances

Grievances Rate
Formula: Number of Employee Grievances / Headcount
Metric Description: Number of Employee Grievances as a percentage of Headcount

% of Grievances Closed
Formula: Number of Employee Grievances Closed / Number of Open Employee Grievances
Metric Description: Employee grievances closed as a percentage of open employee grievances.

5.2.1.3 Increase in employee turnover rate

Employee Turnover Rate
Formula: [Number of employees who left /((Beginning + Ending number of employees)/ 2)] x 100. If you have 1000 employees on 1st January, 1200 employees on 31st December of the same year, and your total employee separations for the year is 100, your company's turnover rate
= 100/((1000+1200)/2))
= 100/1100
= 9%

Metric Description:
Examining trends in turnover rates of your employees can be a good indication of employee engagement, but unfortunately, it's a lagging indicator. Turnover rates within teams will also show you whether there are certain managers or departments that need attention to improve employee experience. Presenteeism can also be an issue. This is where employees arrive to work, but they aren't actively engaged in the activities they are performing. [1]

The term 'employee turnover rate' refers to the percentage of employees who leave an organization during a certain period of time. Depending on what you want to measure, you can use different numbers to calculate your employee turnover rate. For example, if you want to illustrate competitive retention you would normally define separation as voluntary resignations since non-voluntary separations and retirements don't necessarily mean that you're losing employees to other employers. However, if you simply want to illustrate overall turnover, you may want to include all separations. An increase in voluntary turnover can point to a lack of competitiveness in salary, a drop in leadership credibility, poor retention practices or an improved job market.

According to LinkedIn the global average of employee turnover is 12.8%. Turnover rates are drawn from LinkedIn's member data in 2017. It calculates turnover by taking the number of professionals who left their company in a given population, then divides that number by the average amount of people in that given population in 2017. Professionals are considered to have left their jobs if they provide an end-date for their position at a company. (2)

Employee Turnover Rates	
France	21%
United Kingdom	17.6%
Australia	17.5%
Canada	16%
United States	13%
Brazil	10.9%
India	8.7%
South Africa	7.7%

To better understand your employee turnover, answer three questions: (2)
- **Who** are the employees who leave? If your top performers are leaving, take immediate action. On the other hand, if your low performers are leaving, you could stand to gain by enjoying better employee engagement, productivity and profits.
- **When** do they leave? Keeping track of when people leave can be very useful. For example, your new hire turnover rate can offer a lot of insight. First, it can tell you whether your recruitment methods are working. If a large number of your new employees leave because they found their job duties different to, or more complicated than, what they were expecting, perhaps you should consider reviewing your job descriptions.
- **Why** they are leaving? When you know why your employees leave, you can change your company's management style or policies in response.

New Hire Turnover Rate

Formula: (Number of employees who leave after less than 1 year/ number of separations during the same period) x 100

Metric Description: To calculate new employee turnover rate, first determine what period of time you define as new hire turnover (usually when hires leave anywhere before one year of employment.) If your industry has a high employee turnover rate (e.g. hospitality), it might make more sense to track how many employees leave within their first 30, 45, 60 or 90 days, instead of the first year. This metric is one indicator of quality of hire. If this resignation rate is higher for the first year of employment than for employees with a greater length of service, it indicates ineffective recruitment process, on-boarding process or employer branding misalignment.

Data from CEB Global showing new hires are more likely to leave than existing employees

Source: https://fitsmallbusiness.com/internal-vs-external-recruiting/

To reduce turnover rates for new employees:

- **Improve your hiring process**

Hiring the wrong person for a role can cost companies big. Here are some ways to take to reduce employee turnover:
  - **Be clear about what the role entails**. In the job descriptions and during interviews be clear and honest about job requirements, working hours, benefits and salary so that there's no mismatch between expectations and reality.
  - **Evaluate candidates based on job-related criteria**. Assignments and skill assessment tests help you understand if candidates are really good.

- **Plan onboarding programs**

First days at work can set the stage for success.:
  - **Organize employees' first days at work**. Make new-job excitement last by giving new hires a warm welcome. Prepare their workstations, shorten their learning curve with comprehensive onboarding, and schedule team-building activities to help them adjust to their new positions.
  - **Follow up with new hires after 30, 60 and 90 days**. Meet regularly with new hires to find out how they're doing. Assign them a work buddy or mentor for the first few months.

- **Coordinate with hiring managers**

Reducing new hire turnover rates should be a collaborative effort between HR and hiring managers. Advise hiring managers to:
  - **Give new hires meaningful tasks**. Challenge new employees with interesting projects that get them excited and allow them to show off their skills and boost their confidence.
  - **Discuss career paths**. Ask them where they'd like to see themselves in the next few months. New hires will appreciate that you're interested in their career development.

- **Build a healthy workplace**

Consider the following to improve your retention rates:
  - **Build an inclusive culture**. Train managers to become good leaders who'll encourage team members, acknowledge their accomplishments, reject favouritism and bullying and promote open communication.
  - **Offer meaningful perks and benefits**. Consider benefits that make employees more productive such as flexible working hours and work from home. Also, talk to your current employees to learn what kinds of perks would serve as incentives.

Total Top Quartile Performer Resignation Rate
Formula: Number of employees who left in the Top Quartile of Performance / [(Beginning + Ending number of employees)/ 2]
Metric Description: Permanent employees who left and were within the top quartile of performers, as a percentage of average headcount.

Succession Planning Rate
Formula: Executive Candidates / Average Executive Level Headcount
Metric Description: The percentage of permanent executive roles for which there is a succession candidate.

5.2.1.4 Decrease in productivity

Revenue per Headcount
Formula: Revenue / Headcount
Metric Description: The number of dollars of revenue from operations generated per Headcount. This metric can be used to determine if a growing organization should hire. Or, provide a warning when revenue is steady, but the staff required to bring in this revenue is growing at a faster rate.

Profit per Headcount
Formula: (Revenue - Operating Cost) / Headcount
Metric Description: The number of dollars of profit generated per Headcount. It indicates if an organization has the right people, in the right place, at the right price to meet its strategic objectives.

5.2.1.5 Engagement survey participation rates

If the participation rate for an engagement survey is particularly low or high in certain departments or business units, that can be a signal that certain parts of the organization are engaged while others are not. Whether an employee chooses to participate is, in itself, an important piece of feedback.

Reference
(1) Mariya Finkelshteyn (2019) 4 KPIs to Measure Employee Engagement
https://www.saplinghr.com/blog/4-kpis-measure-employee-engagement (1 October 2019)
(2) Canadian HR Reporter (2018) Canada ranks 4th globally for highest employee turnover. https://www.hrreporter.com/focus-areas/culture-and-engagement/canada-ranks-4th-globally-for-highest-employee-turnover/283061 (26 December 2019)

5.2.2 Engagement Dashboards

Well-designed HR dashboards can tell a whole story at a glance. Their data visualizations connect related HR & Business metrics in a clear way to answer questions such as: Did we reach our goals? How far are we from achieving our goals?

5.2.2.1 Engagement dashboard example 1: Sentiment analysis with combination charts

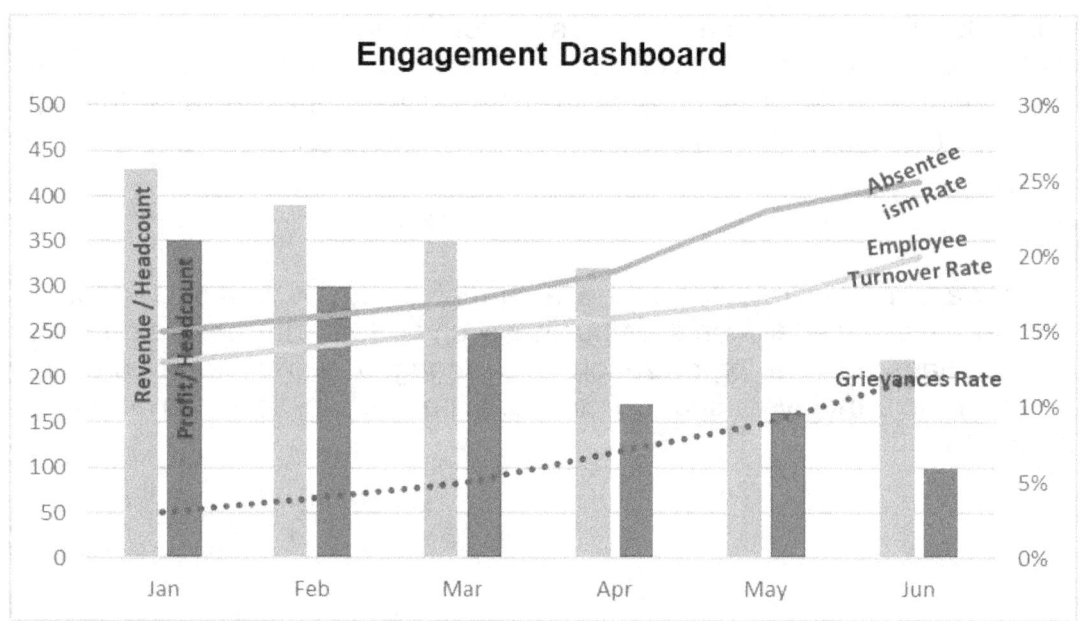

Combination charts can be used to assess employee engagement. The above engagement dashboard show that employees are getting less engaged:
- Increase in absenteeism rate
- Increase in grievances
- Increase in employee turnover rate
- Decrease in productivity (revenue/ headcount)
- Decrease in productivity (profit / headcount)

Steps to create the dashboard in Excel

1) Table below shows an organization's Revenue/headcount, Profit/headcount, Absenteeism rate, Employee Turnover rate, and Grievance rate by months. To better visualize this data, you can create a combination chart with Excel. Select the range A2:F8.

	A	B	C	D	E	F
1	Engagement Dashboard: Sentimal Analysis					
2	Month	Revenue / Headcount	Profit/ Headcount	Absenteeism Rate	Employee Turnover Rate	Grievances Rate
3	Jan	430	350	15%	13.0%	3.0%
4	Feb	390	300	16%	14.0%	4.0%
5	Mar	350	250	17%	15.0%	5.0%
6	Apr	320	170	19%	16.0%	7.0%
7	May	250	160	23%	17.0%	9.0%
8	Jun	220	100	25%	20.0%	12.0%

2) Go to the Inset tab > Charts group, click the "Combo" chart icon, and then click "Create Custom Combo Chart".

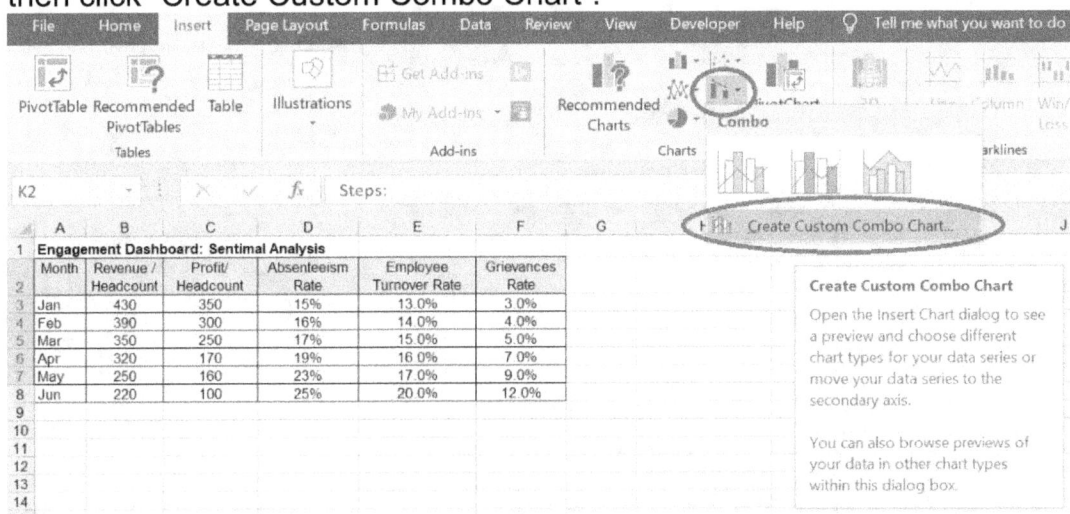

3) The "Insert Chart" dialog box appears.
- Choose "Clustered Column" chart type for:
  - Revenue/Headcount
  - Profit/Headcount series,
- Choose "Line" chart type and check the "secondary axis" for:
  - Absenteeism Rate
  - Employee Turnover Rate
  - Grievances Rate series,
- Click "OK"

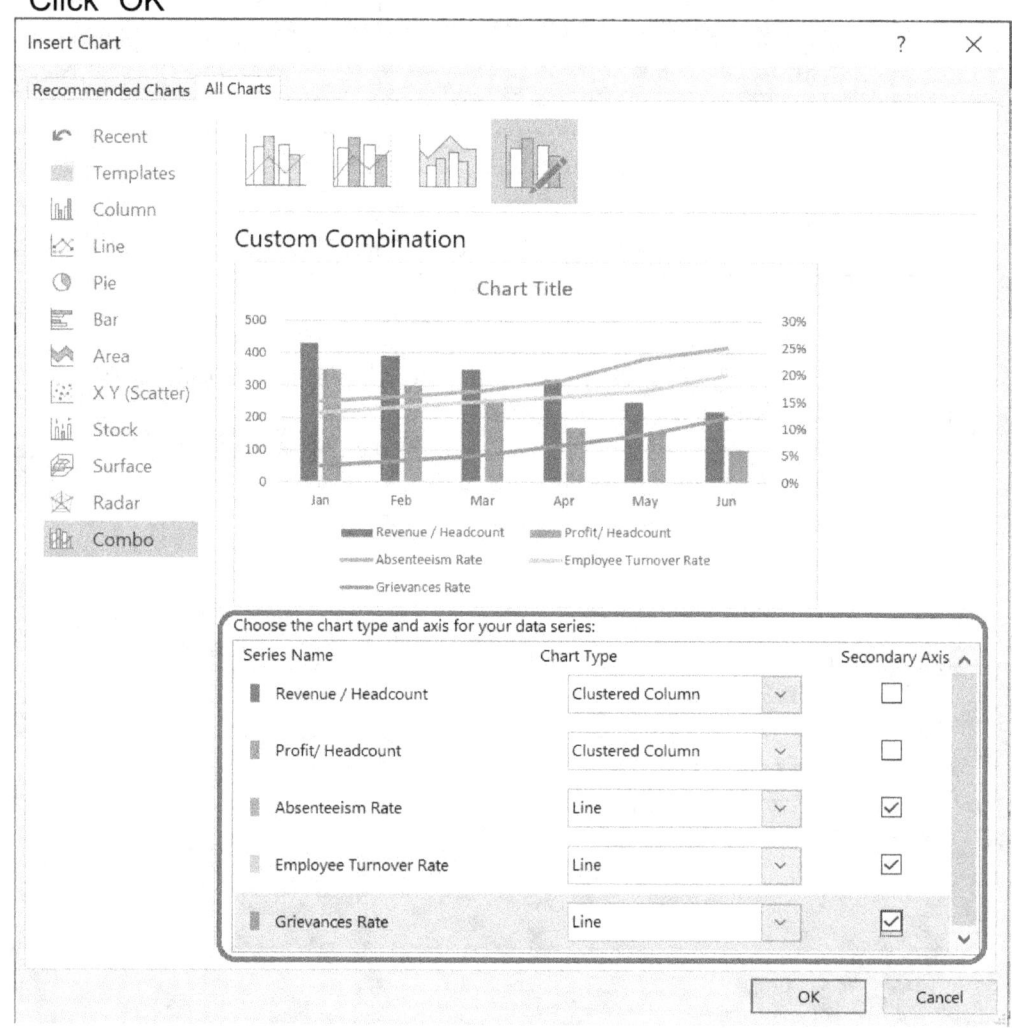

4) The combination chart will be immediately inserted in your worksheet. Left click on the word "Chart Title" to rename it to "Engagement Dashboard".

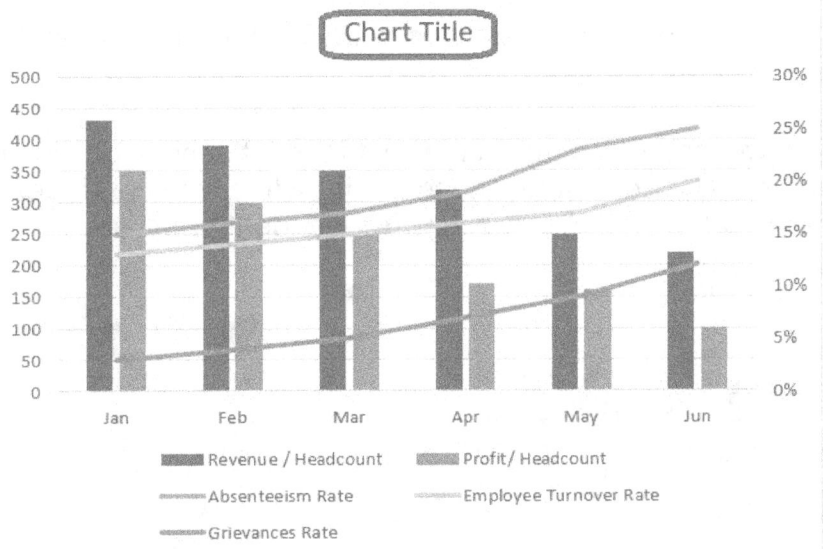

5) After you rename on the word "Chart Title" to "Engagement Dashboard", you will get the below chart.

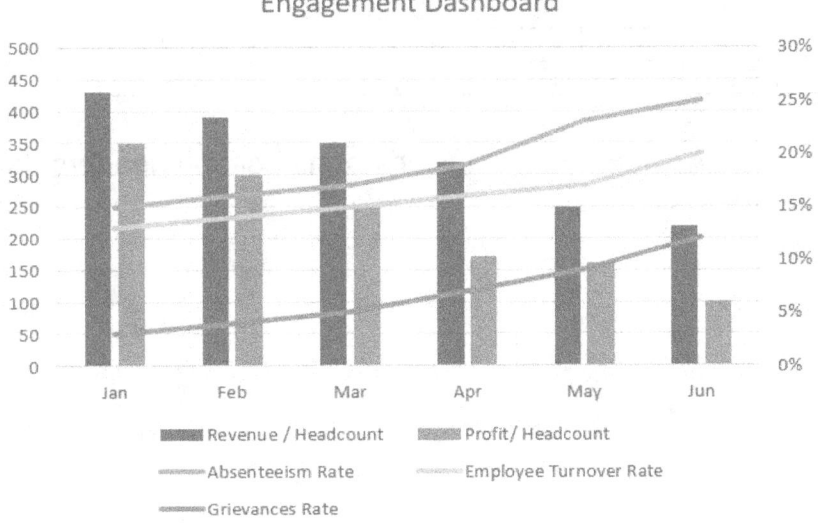

5.2.2.2 Engagement dashboard example 2: Identify predictors of profitability with scatter plots

Scatter plots can be used to visualize how columns of data are related to each other.

1) In our example, we are going to visualize the relationship between "Employee Engagement Index" and "Leadership Effectiveness Index" (independent variable) with "Profit/Headcount" (dependent variable). First, arrange your data.

	A	B	C	D
1	Engagement Dashboard: Identify predictors of profitability with scatter plot			
2	Year	Employee Engagement Index (EEI)	Leadership Effectiveness Index (LEI)	Profit/ Headcount
3	Year 1	4.30	6.30	$100
4	Year 2	4.60	5.90	$135
5	Year 3	4.20	6.40	$120
6	Year 4	5.30	6.60	$140
7	Year 5	5.50	6.70	$130
8	Year 6	5.80	7.10	$150
9	Year 7	5.40	7.30	$160
10	Year 8	6.30	7.80	$170
11	Year 9	6.80	8.50	$180
12	Year 10	7.50	8.70	$350

2) Select two columns with numeric data, including the column headers. In our case, it is the range B2:D12.

3) Go to the Inset tab > Charts group, click the "Combo" chart icon, and then click "Create Custom Combo Chart".

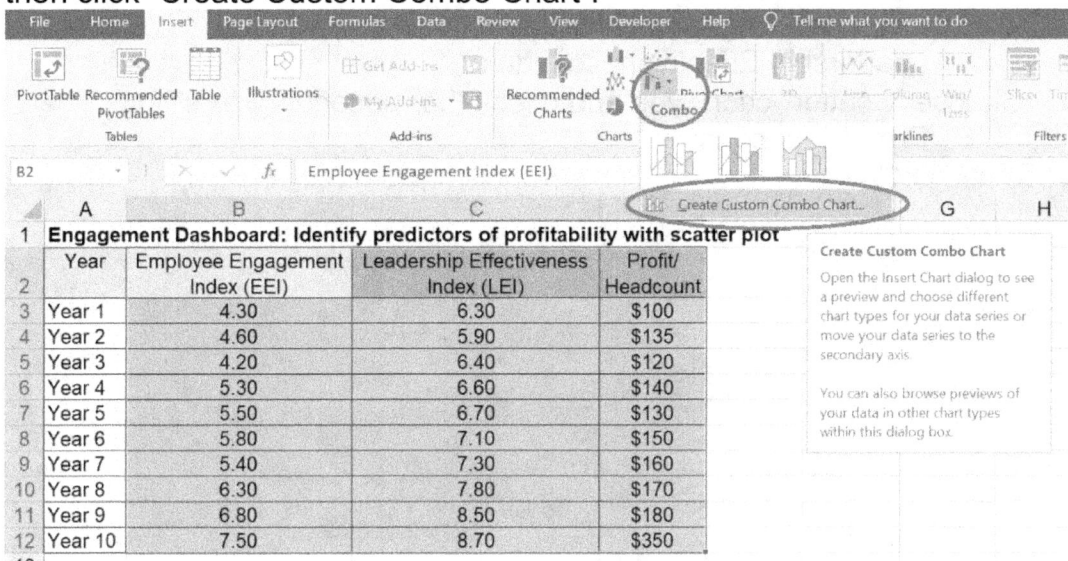

4) The "Insert Chart" dialog box appears.
- Choose "Scatter" chart type for:
    - Employee Engagement Index
    - Leadership Effectiveness Index
- Choose "Scatter" chart type and check the "secondary axis" for:
    - Profit/Headcount
- Click "OK"

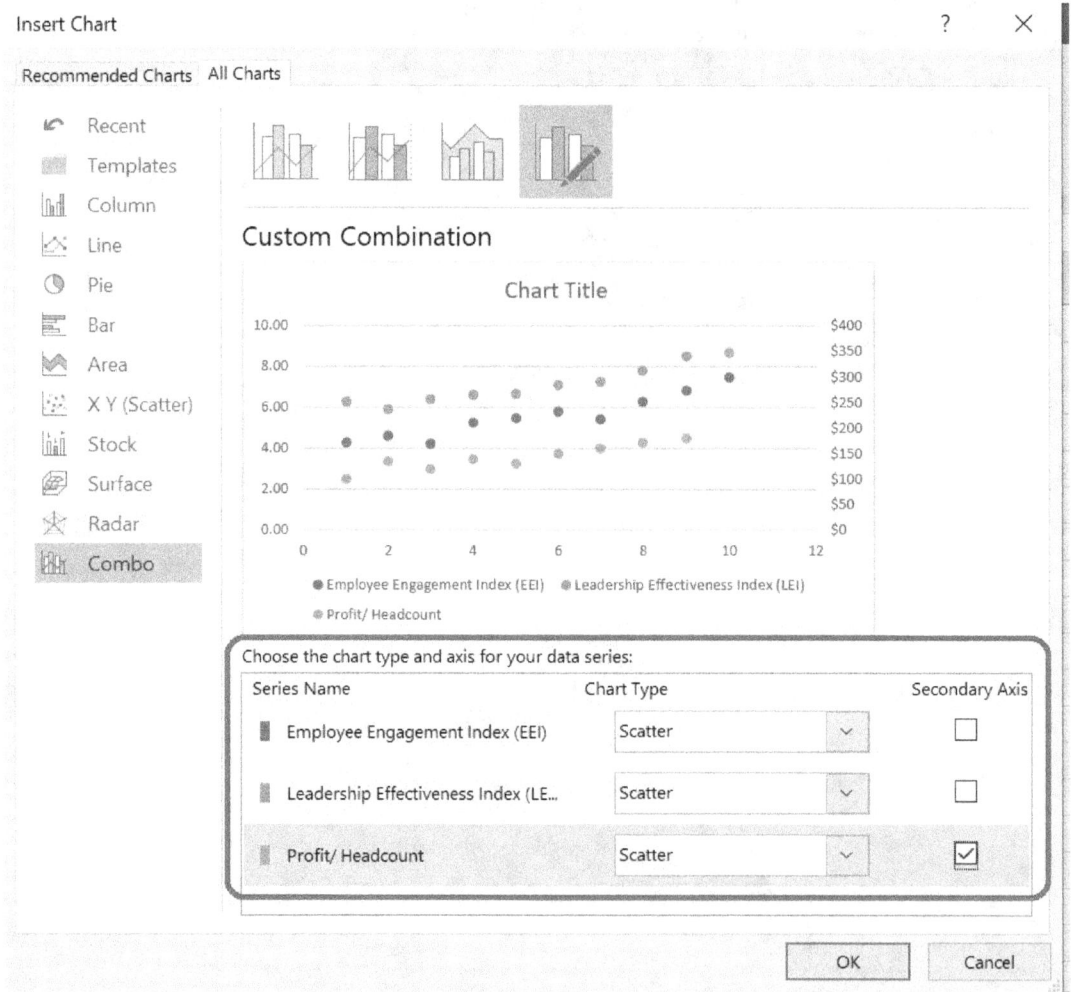

5) The combination chart will be immediately inserted in your worksheet. Left click on the word "Chart Title" to rename it to "Engagement Dashboard".

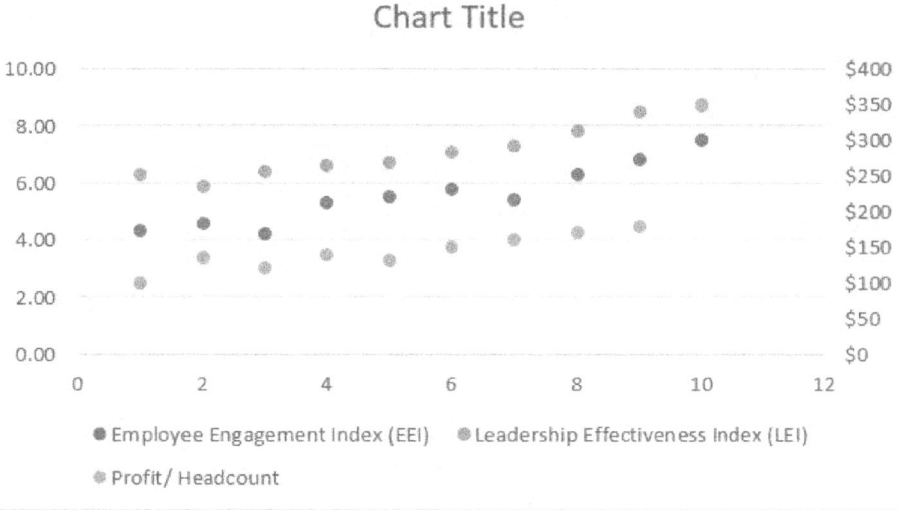

6) After you rename on the word "Chart Title" to "Engagement Dashboard", you will get the below chart.

7) To better visualize the relationship between the two variables, you can draw a trendline in your scatter graph (also called a line of best fit). Right click on any data point and choose "Add Trendline" from the menu.

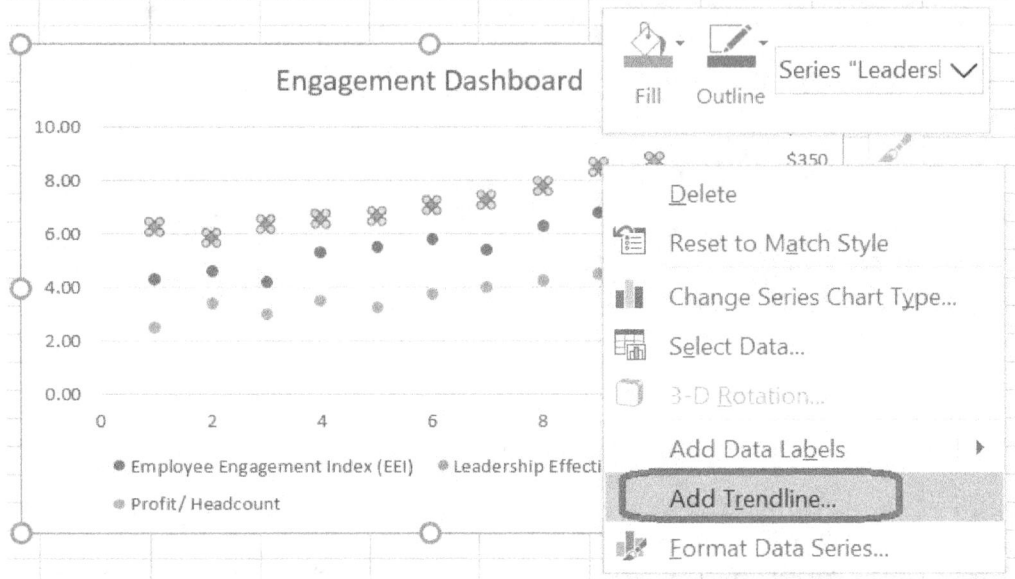

8) Check the "Linear" trendline option. Excel will draw a line as close as possible to all data points so that there are as many points above the line as below.

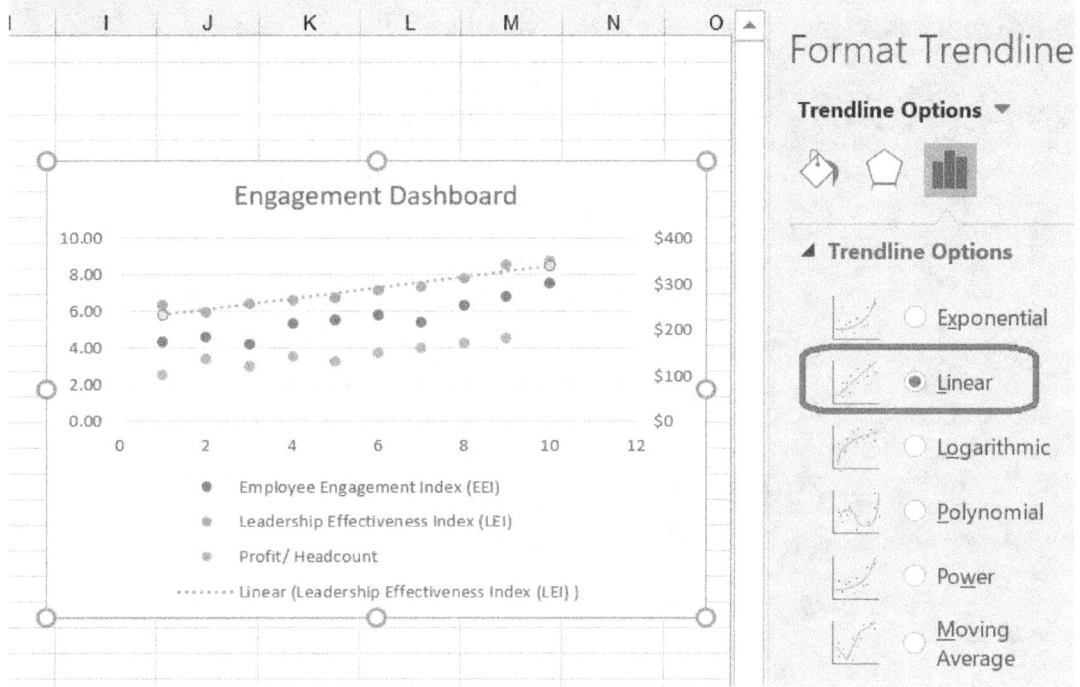

9) Repeat step 5 and 6 to create the trendlines for the other data sets.

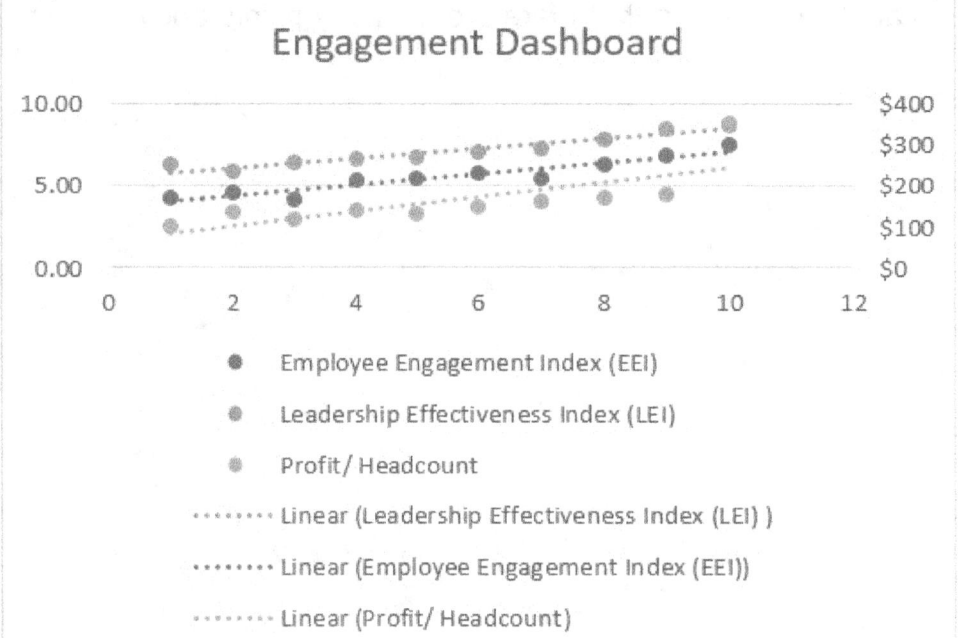

5.2.2.3 Engagement dashboard example 3: Predict profitability with scatter plot & trendline equation

A scatter plot (also called XY graph) is a two-dimensional chart that shows the relationship between variables. Usually, the independent variable is on the x-axis, and the dependent variable on the y-axis. The values are displayed at data points at the x and y axis intersection. The scatter plot shows how strength of correlation between two variables. The closer data points are along a straight line, the stronger the correlation.

From the chart below, we conclude that the relationship between the two variables are linear. i.e. as the value of x variable increases, there is a corresponding increase in the value of the y variable.

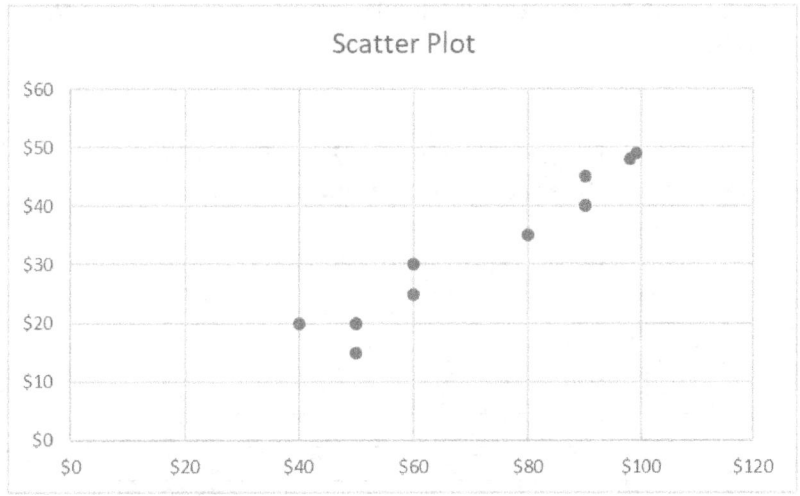

When to use a scatter plot:
- When you want to analyze and report the relationship/correlation between two variables.
- When there are at least 10 data points. The more the data points the better.

In this section, you will learn how to do a scatter plot in Excel to create a graphical representation of two correlated data sets. If you have to analyse the data below to identify patterns or trends, it would be a tedious task. However, a scatter plot can help you to visualize how columns of data are related to each other.

Steps to create a scatter plot in Excel

1) In our example, we are going to visualize the relationship between "Employee Engagement Index" (independent variable) and "Profit/Headcount" (dependent variable). First, arrange your data.
- **Independent variable** should be in the left column as it will be plotted on the x axis.
- **Dependent variable** should be in the right column, as it will be plotted on the y axis.

	A	B	C
1		Engagement Dashboard: Predict profitability with scatter plot & trendline equation	
2	Month	Employee Engagement Index (EEI)	Profit/Headcount
3	Jan	4.30	$100
4	Feb	4.60	$135
5	Mar	4.20	$120
6	Apr	5.30	$140
7	May	5.50	$130
8	Jun	5.80	$150
9	Jul	5.40	$160
10	Aug	6.30	$170
11	Sep	6.80	$250
12	Oct	6.30	$300
13	Nov	7.00	$350
14	Dec	7.50	$370

2) Select two columns with numeric data, including the column headers. In our case, it is the range B2:C14.

3) Go to the Inset tab > Charts group, click the "scatter" chart icon.

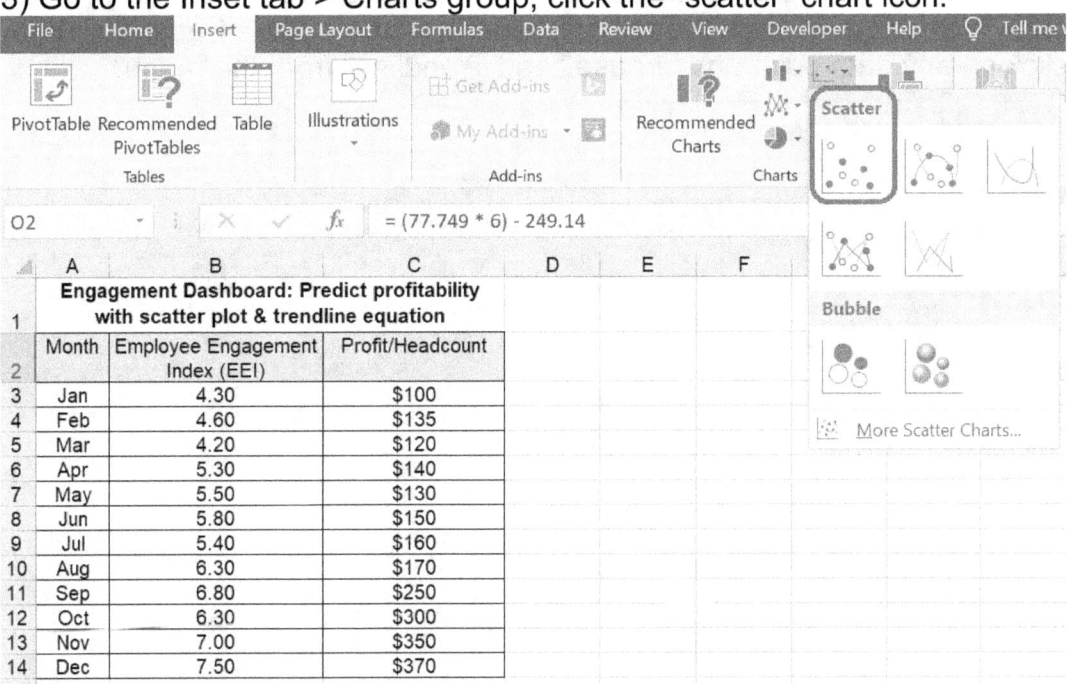

4) The scatter diagram will be immediately inserted in your worksheet:

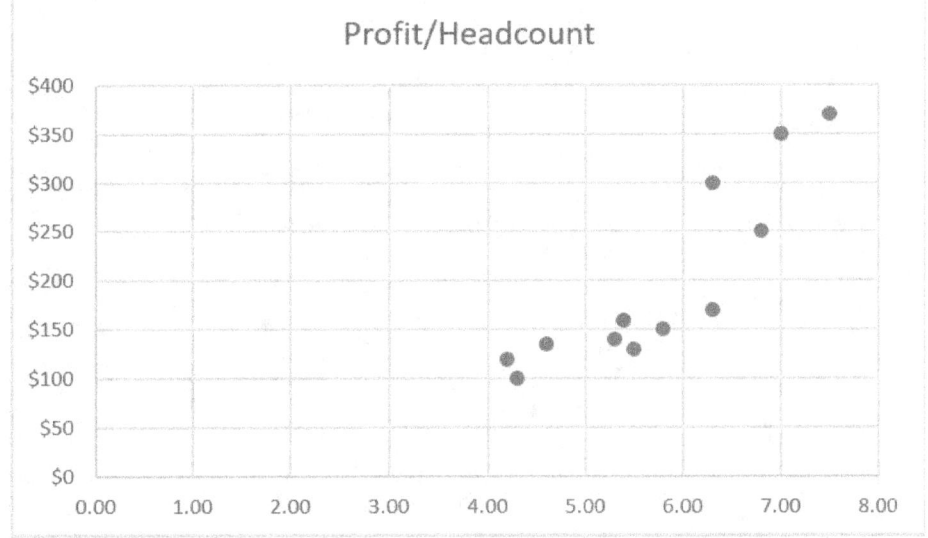

5) To better visualize the relationship between the two variables, you can draw a trendline in your scatter graph (also called a line of best fit). Right click on any data point and choose "Add Trendline" from the menu.

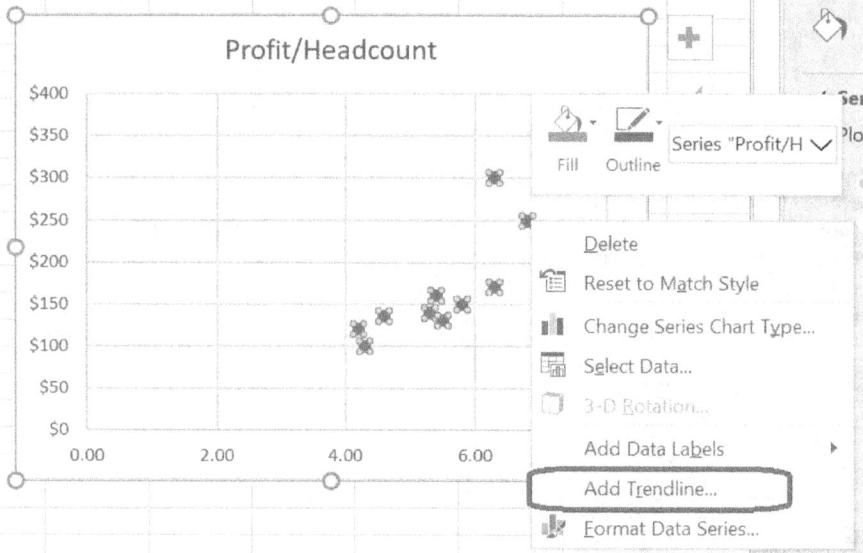

6) Check the "Linear" trendline option. Excel will draw a line as close as possible to all data points so that there are as many points above the line as below.

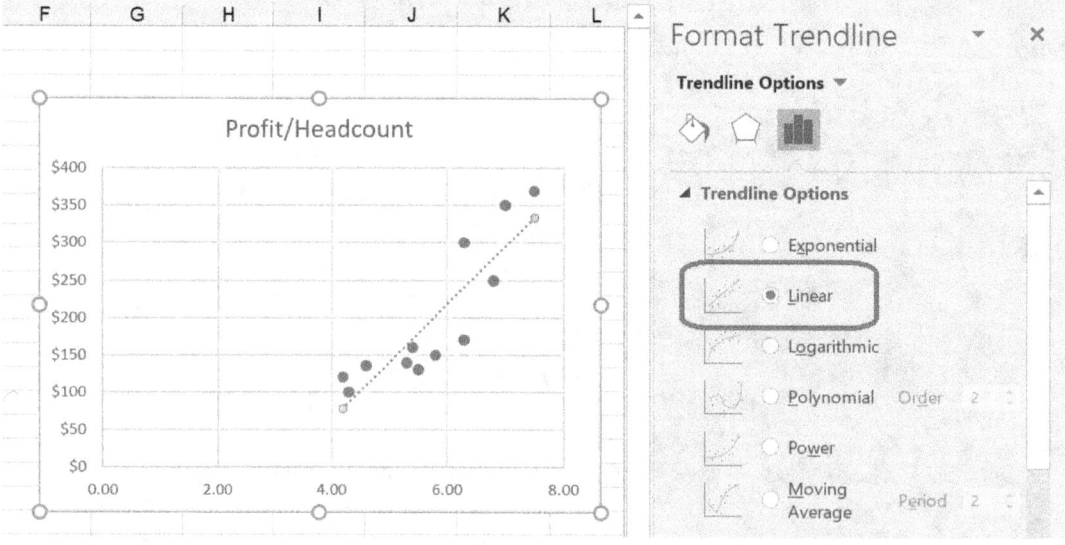

7) Check the "Display Equation on chart" the box - to show the equation for the trendline that mathematically describes the relationship between the two variables. Check the "Display R-squared value on chart" – to show the fit (near to 1 means it is a very good fit).

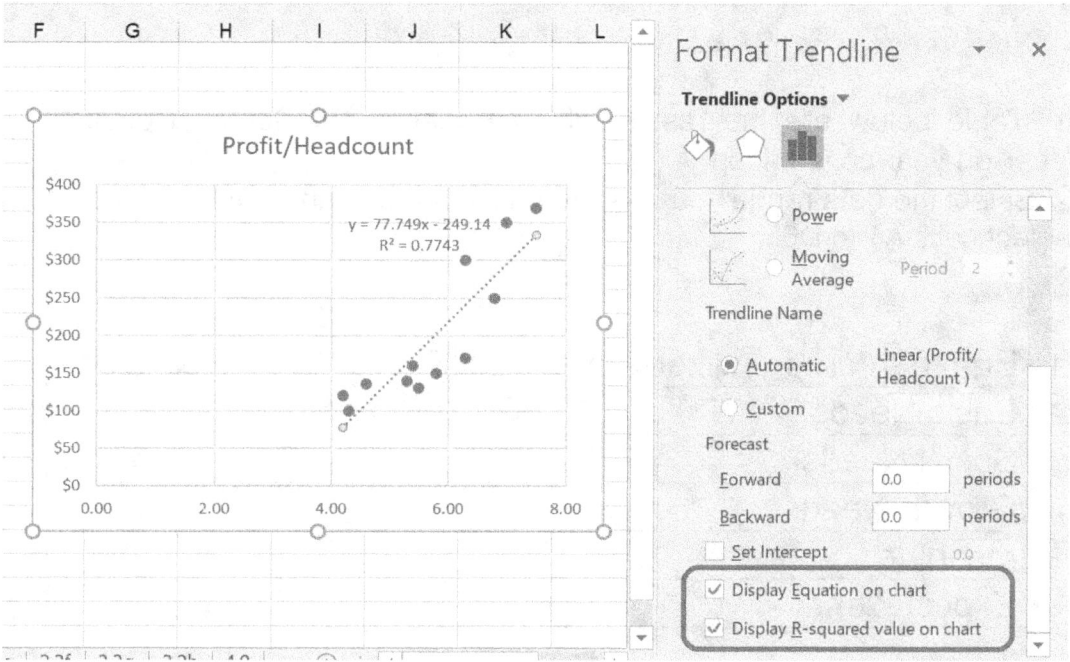

From the Correlation Output, the R^2 is 0.7743 (near to 1 means it is a very good fit).

From the Correlation Output, the correlation equation is:
y = 77.749x - 249.14

Thus, if the Engagement Score is 6, the predicted Profit/Headcount
= 77.749x - 249.14
= (77.749 x 6) - 249.14
= **217**

5.3 Bar Charts

These are several ways to analyse the engagement survey results:
- Year-on-year comparison
- Business unit vs Market average vs Global average

(i) Year-on-year comparison

1. Table below are the results of an Engagement Survey Results for: Year-on-year comparison.
2. Select the data range you need to show in the chart. In this example select cells A1 to D6.

	A	B	C	D
1	Engagement Fertilizers	Year 2017	Year 2018	Year 2019
2	Basic Needs	6	5	4
3	Social Needs	7	6	5
4	Growth Needs	5	4	3
5	Meaning	6	5	4
6	Expectations	6	5	4

3. Click **Insert** > **Insert Column or Bar Chart**, and select **Clustered Column**. See screenshot:

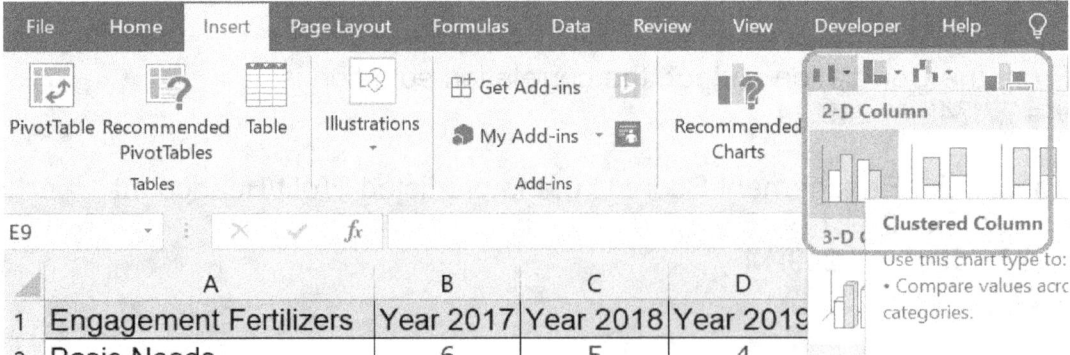

4. Now the radar chart is created with the axis labels.

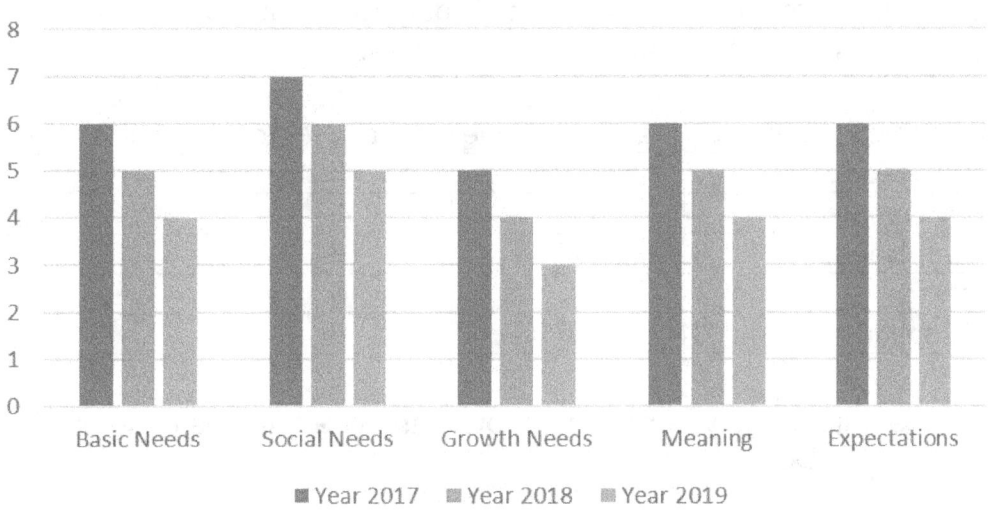

5. To edit chart title, click the chart title that you want to change, and type the new title as "Engagement Survey Results (Year-on-year comparison)".

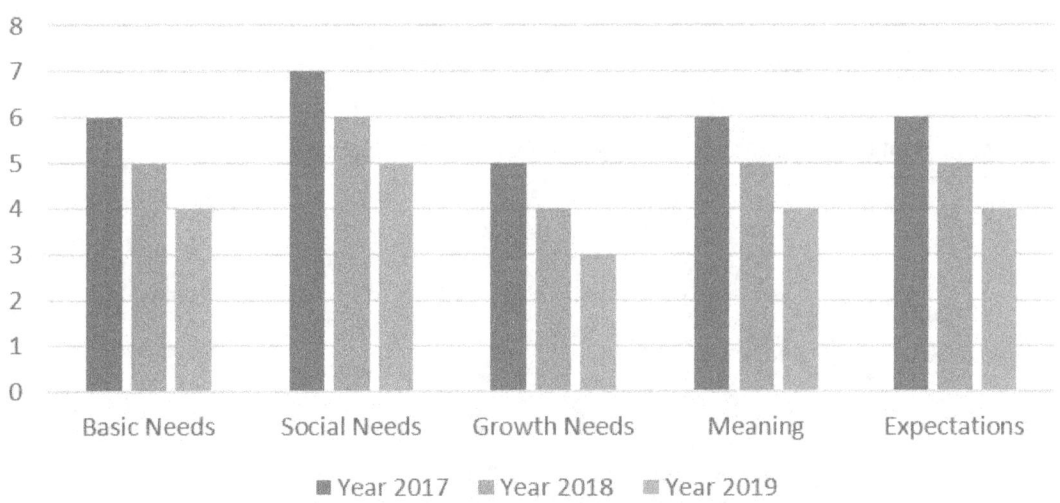

(ii) Business unit vs Market average vs Global average

1. Table below are the results of an Engagement Survey Results for: Business unit vs Market average vs Global average.

	A	B	C	D
1	Engagement Fertilizers	Market average	Global average	Business unit
2	Basic Needs	6	5	4
3	Social Needs	7	6	5
4	Growth Needs	5	4	3
5	Meaning	6	5	4
6	Expectations	6	5	4

2. Select the data range you need to show in the chart. In this example select cells A1 to D6.

3. Click **Insert** > **Insert Column or Bar Chart**, and select **Clustered Column**. See screenshot:

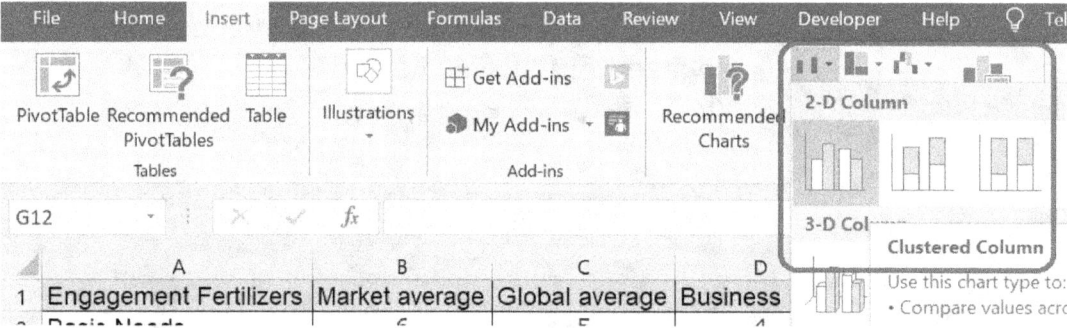

4. Now the radar chart is created with the axis labels.

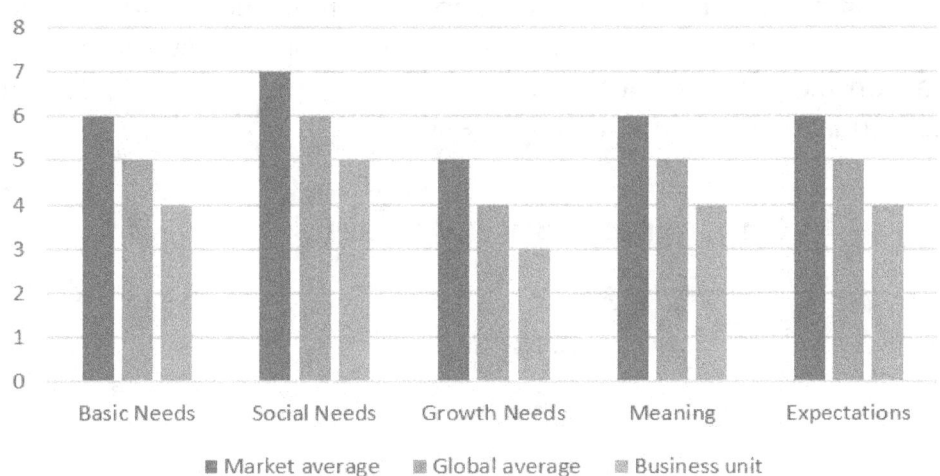

5. To edit chart title, click the chart title that you want to change, and type the new title as "Engagement Survey Results (Business unit vs Market average vs Global average)".

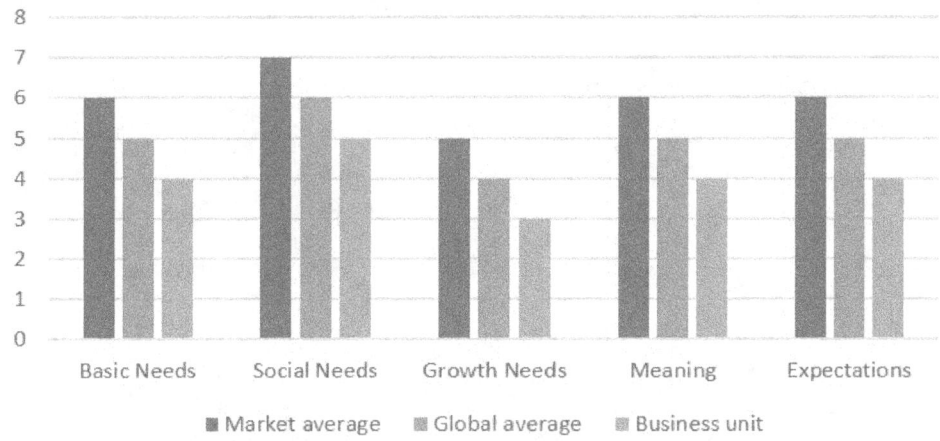

5.4 Radar Charts

A radar chart is a graphical method of displaying multivariate data in the form of a two-dimensional chart of three or more quantitative variables represented on axes starting from the same point. The Radar Chart can be used to visualize engagement survey results.

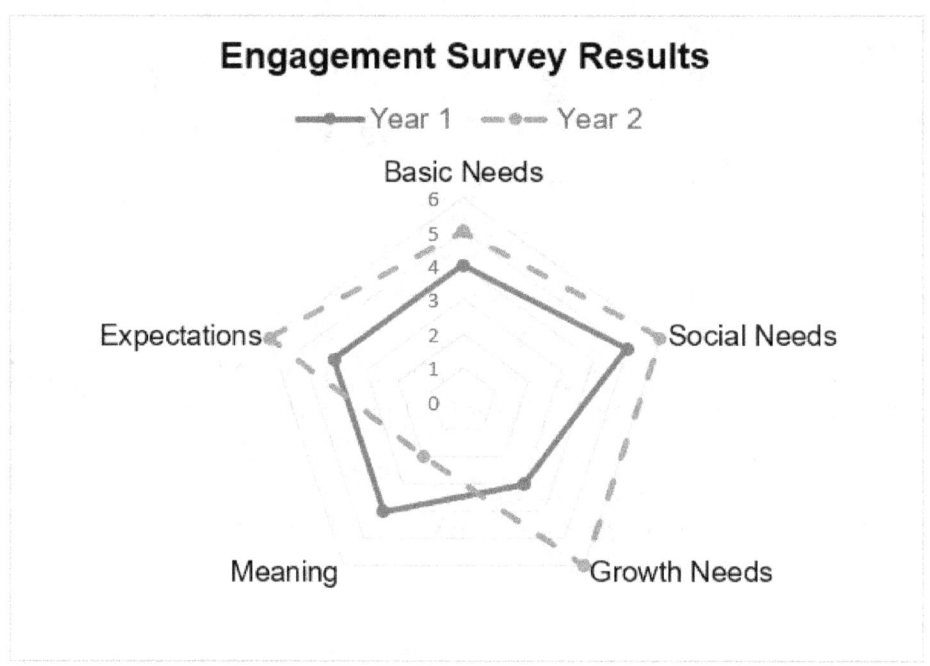

1) Table below shows the ratings for the results of an engagement survey results. Select the data range you need to show in the chart. In this example select cells A1 to C15. See screenshot:

	A	B	C
1	Engagement Fertilizers	Year 1	Year 2
2	Basic Needs	4	5
3	Social Needs	5	6
4	Growth Needs	3	6
5	Meaning	4	2
6	Expectations	4	6

2. Click **Insert** > **Waterfall** > **Radar**, and select **Radar with Markers**. See screenshot:

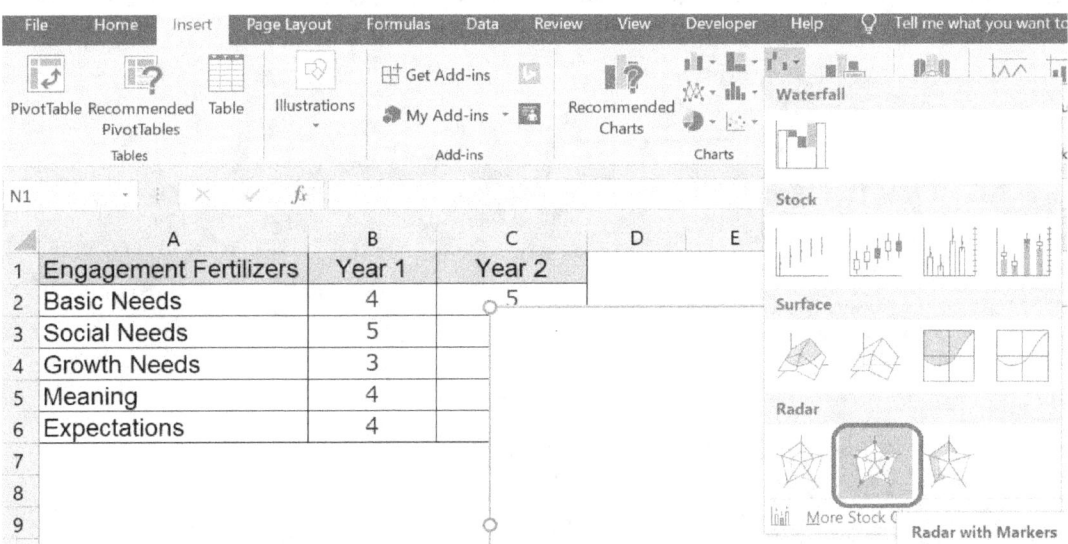

3. Now the radar chart is created with the axis labels.

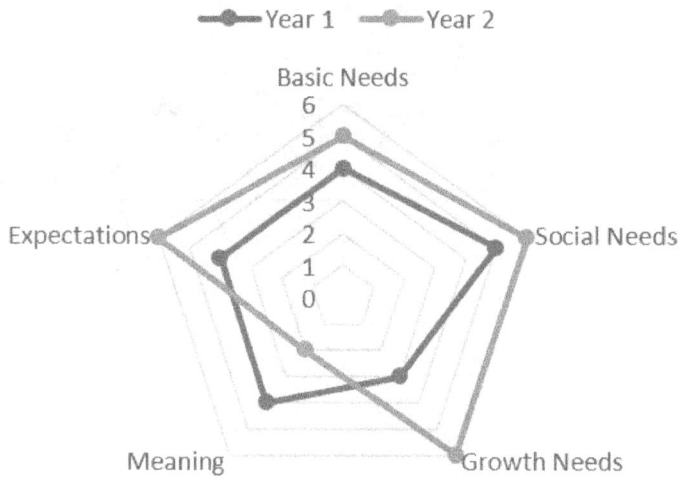

4. To change the "Year 2" line from solid to dotted, right click on the "Year 2" line, then left click "Format Data Series".

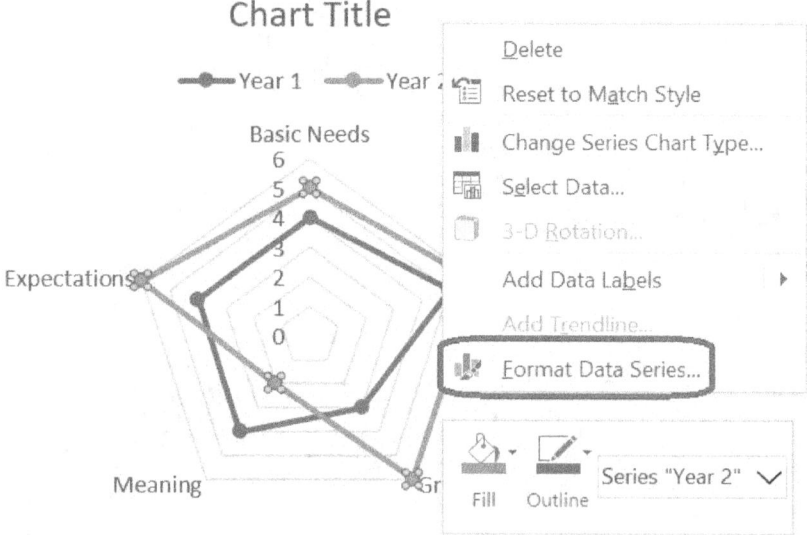

5. At the "Format Chart Area" panel on the right, click the "Fill & Line" tab.

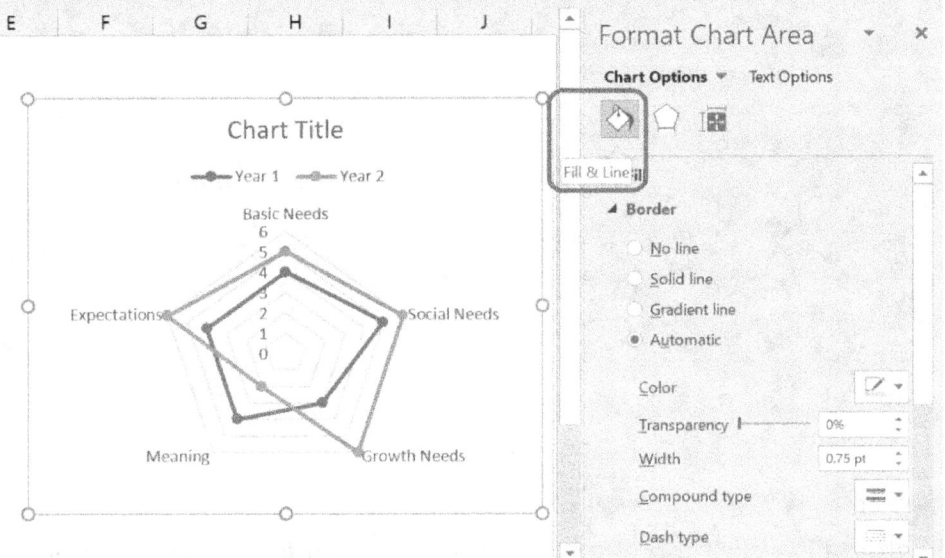

6. Click **Dash type** and select **Dash.**

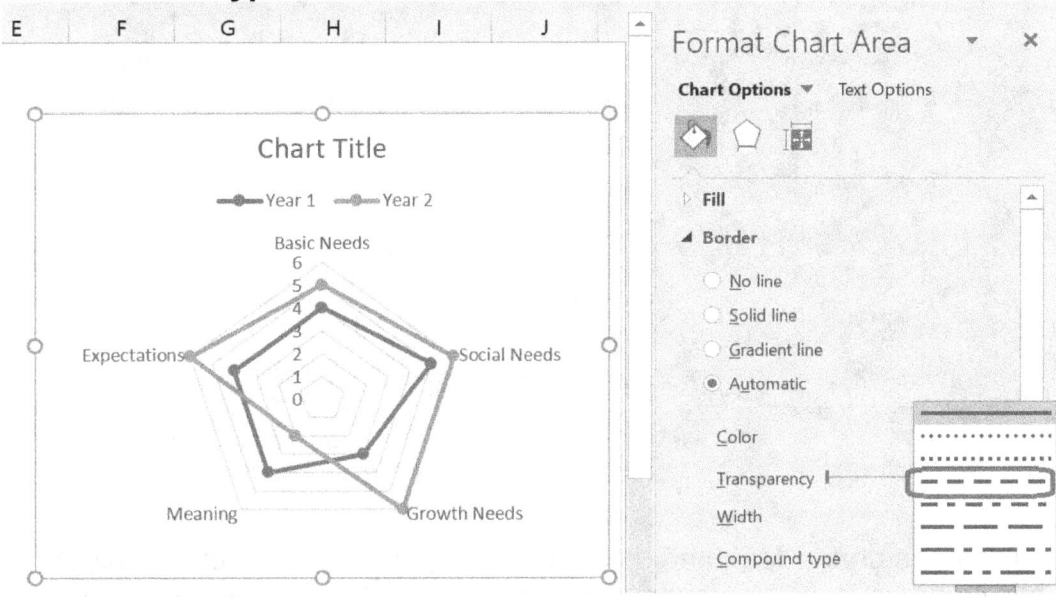

7. Now the "Year 2" line is changed from solid to dotted

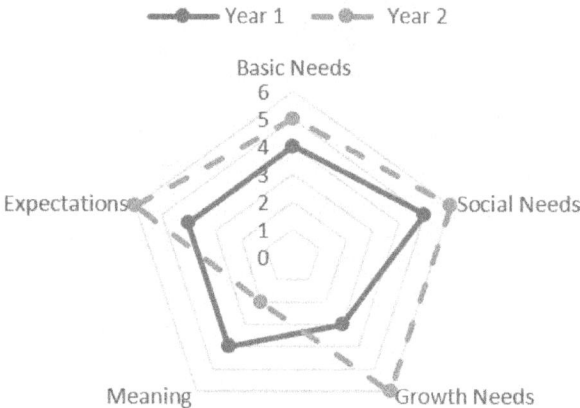

8. To edit chart title, click the chart title that you want to change, and type the new title as "Engagement Survey Results".

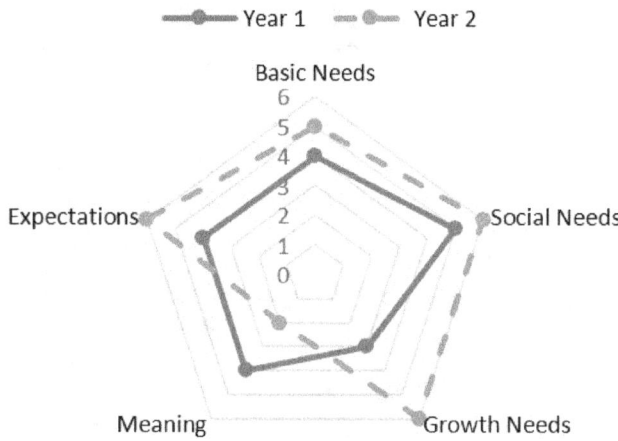

9. From the chart, "Meaning" is the only Engagement Survey Result that is higher in Year 1 compared to Year 2. The other Engagement Survey Results (Basic Needs, Social Needs, Growth Needs, Expectations) are higher in Year 2 than Year 1.

5.5 Correlation analysis of engagement prescription

In this example, we use correlation to find out whether there is a relationship between "IISS model's 5 Engagement Fertilizers score" and "Revenue per headcount"

	A	B	C	D	E	F	G
1		\multicolumn{5}{c	}{IISS model's 5 Engagement Fertilizers Score}		
2	Year	Basic Needs Score	Social Needs Score	Growth Needs Score	Meaning Score	Expectations Score	Revenue per headcount
3	1	8.3	7.8	9.6	8.8	9.5	1600
4	2	8.5	7.5	8.9	8.3	9.2	1500
5	3	6.8	7.2	9.3	8.2	8.7	1400
6	4	6.5	6.8	9.2	7.9	8.3	1300
7	5	6.2	6.7	9.1	7.4	7.9	1200
8	6	5.9	6.3	8.9	7.3	7.3	1100
9	7	5.7	6.1	8.8	7.1	6.9	1000
10	8	5.5	5.8	8.7	6.8	6.4	900
11	9	5.3	5.6	8.6	6.5	5.8	800
12	10	4.5	5.3	6.2	6.2	5.5	700

1) Install "Analysis ToolPak", an Excel add-in

"Analysis ToolPak" is an add-in for Microsoft Excel that comes with Microsoft Excel. To be able to run regression using Excel, you need to first install "Analysis ToolPak", an Excel add-in program that provides data analysis tools. To load the Analysis ToolPak add-in, follow these steps:

- On the File tab, click Options.

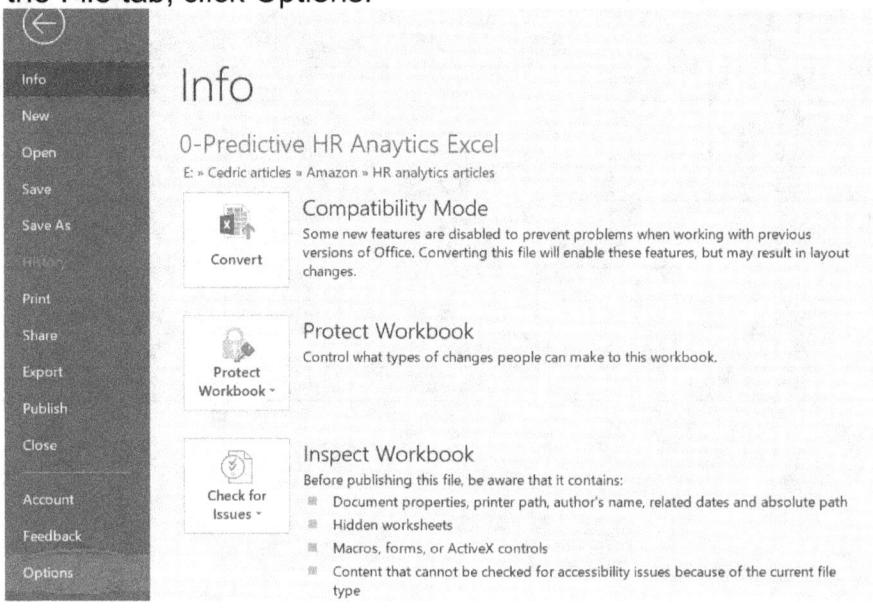

- Under Add-ins, click Analysis ToolPak and click the "Go" button.

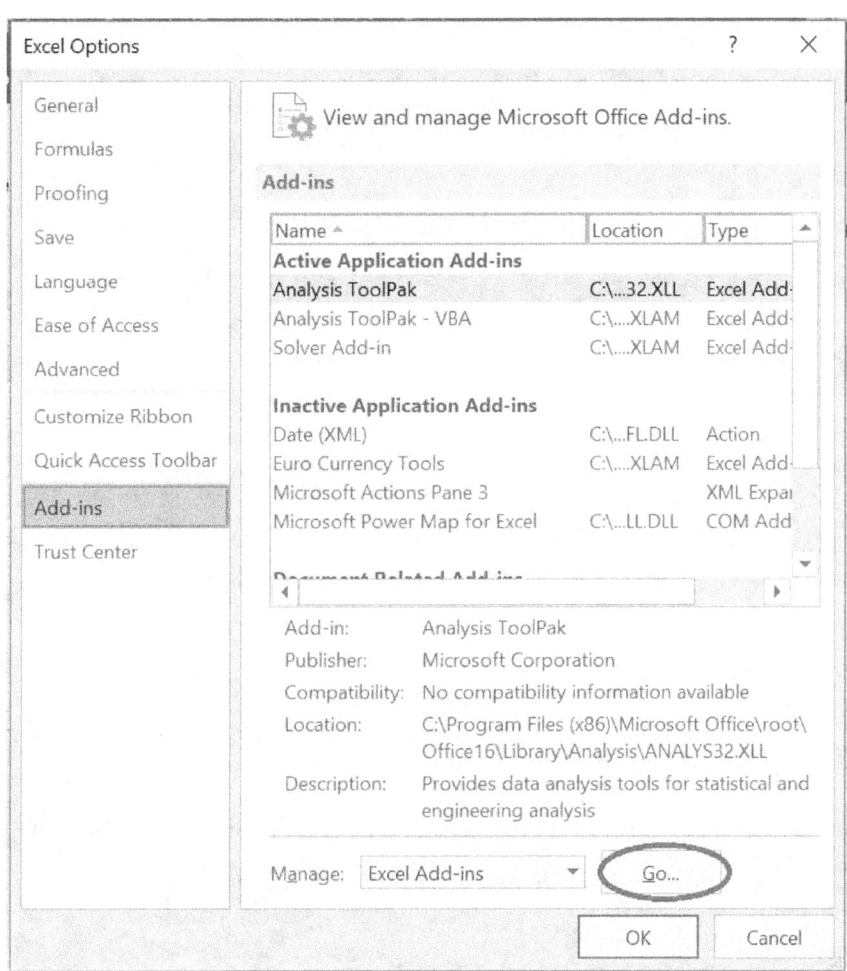

- Click "Analysis ToolPak" and click on OK.

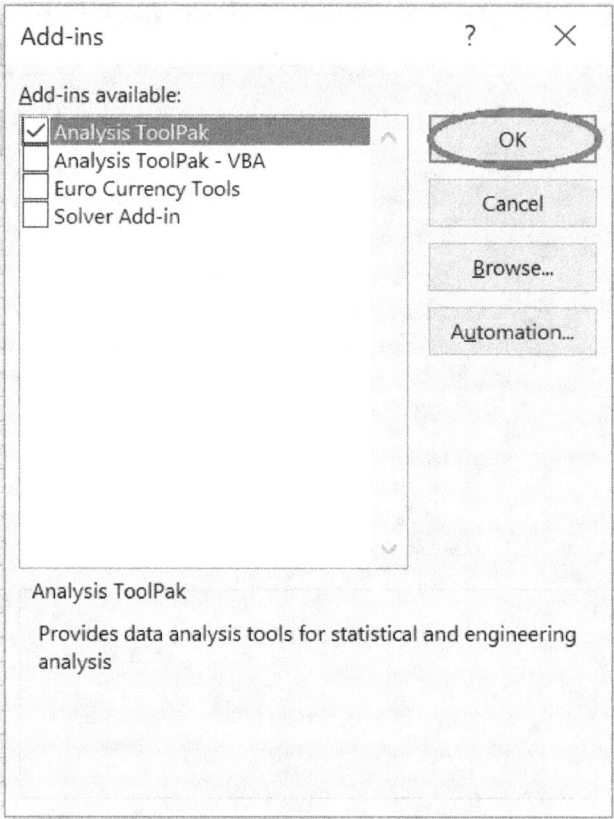

- On the Data tab, in the Analysis group, you are now able to click on "Data Analysis".

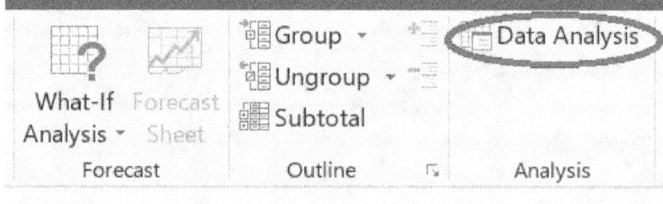

2) Copy the example data in the following table, and paste it in cell A1 of a new Excel worksheet.

	A	B	C	D	E	F	G
1		IISS model's 5 Engagement Fertilizers Score					
2	Year	Basic Needs Score	Social Needs Score	Growth Needs Score	Meaning Score	Expectations Score	Revenue per headcount
3	1	8.3	7.8	9.6	8.8	9.5	1600
4	2	8.5	7.5	8.9	8.3	9.2	1500
5	3	6.8	7.2	9.3	8.2	8.7	1400
6	4	6.5	6.8	9.2	7.9	8.3	1300
7	5	6.2	6.7	9.1	7.4	7.9	1200
8	6	5.9	6.3	8.9	7.3	7.3	1100
9	7	5.7	6.1	8.8	7.1	6.9	1000
10	8	5.5	5.8	8.7	6.8	6.4	900
11	9	5.3	5.6	8.6	6.5	5.8	800
12	10	4.5	5.3	6.2	6.2	5.5	700

3) Select "Correlation" and click "OK".

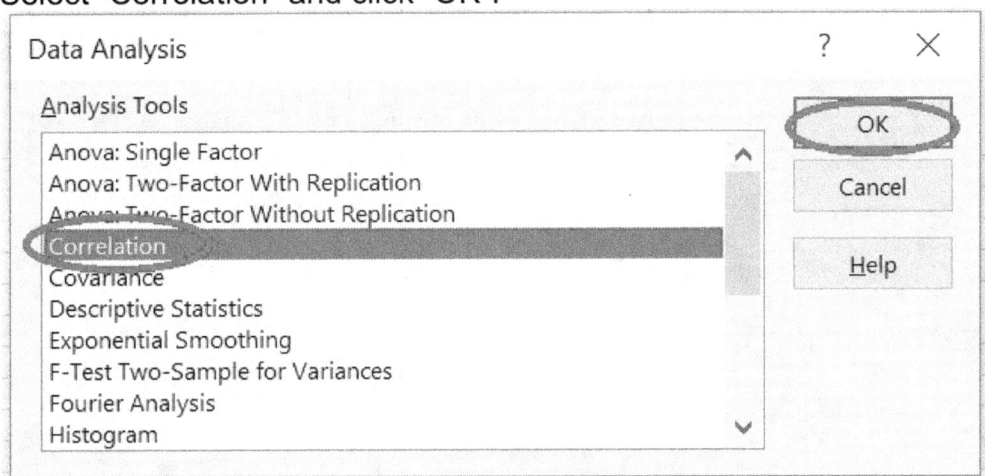

4) After you click OK in the "Data Analysis" dialog box, you will see a "Correlation" dialog box.
5) For "Input Range", select cells (B2:G12).
6) Check "Labels in first row"
7) For "Output Range", select cells (A14).
8) Click "OK"

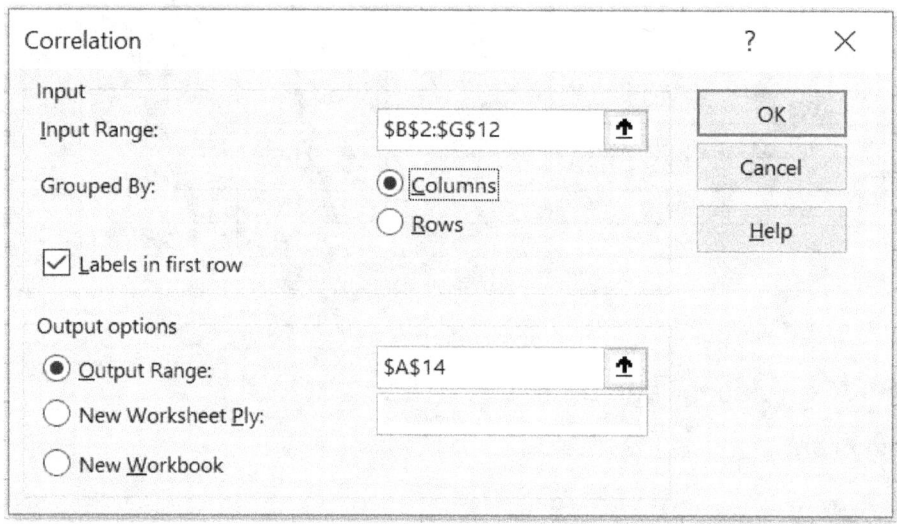

9) After you click "OK", Excel generates the following Correlation analysis.

	Basic Needs Score	Social Needs Score	Growth Needs Score	Meaning Score	Expectations Score	Revenue per headcount
Basic Needs Score	1					
Social Needs Score	0.96	1				
Growth Needs Score	0.66	0.71	1			
Meaning Score	0.94	0.99	0.73	1		
Expectations Score	0.94	0.99	0.72	0.99	1	
Revenue per headcount	0.95	1.00	0.73	0.99	1.00	1

A negative correlation coefficient means that an increase in X is associated with a decrease in Y. Similar to a positive correlation, a negative correlation shows a connection between two variables, and the relative strengths are the same. In other words, a correlation coefficient of 0.85 has the same strength as a correlation coefficient of -0.85. Correlation coefficients are always values between -1 and 1, where "-1" means that there is a perfect linear negative correlation, while "1" shows a perfect linear positive correlation. A correlation coefficient of zero, or near to zero, means that there is no meaningful relationship between variables. Correlation coefficient of 0.91 or -0.92 shows a very strong positive and negative correlation respectively. However, correlation does not mean causation.

An example of negative correlation is the amount of snowfall and the temperature. As the temperature increases, the amount of snowfall decreases. An example of positive correlation is the relationship between temperature and ice cream sales. As temperature increases, so do ice cream sales.

10) From the Excel Correlation analysis,
- **Basic Needs Score**: correlation of 0.95 with Revenue per headcount
- **Social Needs Score:** correlation of 1.00 with Revenue per headcount
- **Growth Needs Score:** correlation of 0.73 with Revenue per headcount
- **Meaning Score:** correlation of 0.99 with Revenue per headcount
- **Expectations Score:** correlation of 1.00 with Revenue per headcount

"Expectations Score" and "Social Needs Score" has the highest correlation with Revenue per headcount.

	Basic Needs Score	Social Needs Score	Growth Needs Score	Meaning Score	Expectations Score	Revenue per headcount
Basic Needs Score	1					
Social Needs Score	0.96	1				
Growth Needs Score	0.66	0.71	1			
Meaning Score	0.94	0.99	0.73	1		
Expectations Score	0.94	0.99	0.72	0.99	1	
Revenue per headcount	0.95	1.00	0.73	0.99	1.00	1

5.6 Regression analysis of engagement prescription

In this example, we use regression to predict the impact of changes in the "IISS model's 5 Engagement Fertilizers score" on "Revenue per headcount"

	A	B	C	D	E	F	G
1		\multicolumn{5}{c}{IISS model's 5 Engagement Fertilizers Score}	
2	Year	Basic Needs Score	Social Needs Score	Growth Needs Score	Meaning Score	Expectations Score	Revenue per headcount
3	1	8.3	7.8	9.6	8.8	9.5	1600
4	2	8.5	7.5	8.9	8.3	9.2	1500
5	3	6.8	7.2	9.3	8.2	8.7	1400
6	4	6.5	6.8	9.2	7.9	8.3	1300
7	5	6.2	6.7	9.1	7.4	7.9	1200
8	6	5.9	6.3	8.9	7.3	7.3	1100
9	7	5.7	6.1	8.8	7.1	6.9	1000
10	8	5.5	5.8	8.7	6.8	6.4	900
11	9	5.3	5.6	8.6	6.5	5.8	800
12	10	4.5	5.3	6.2	6.2	5.5	700

1) Install "Analysis ToolPak", an Excel add-in

"Analysis ToolPak" is an add-in for Microsoft Excel that comes with Microsoft Excel. To be able to run regression using Excel, you need to first install "Analysis ToolPak", an Excel add-in program that provides data analysis tools. To load the Analysis ToolPak add-in, follow these steps:

On the File tab, click Options.

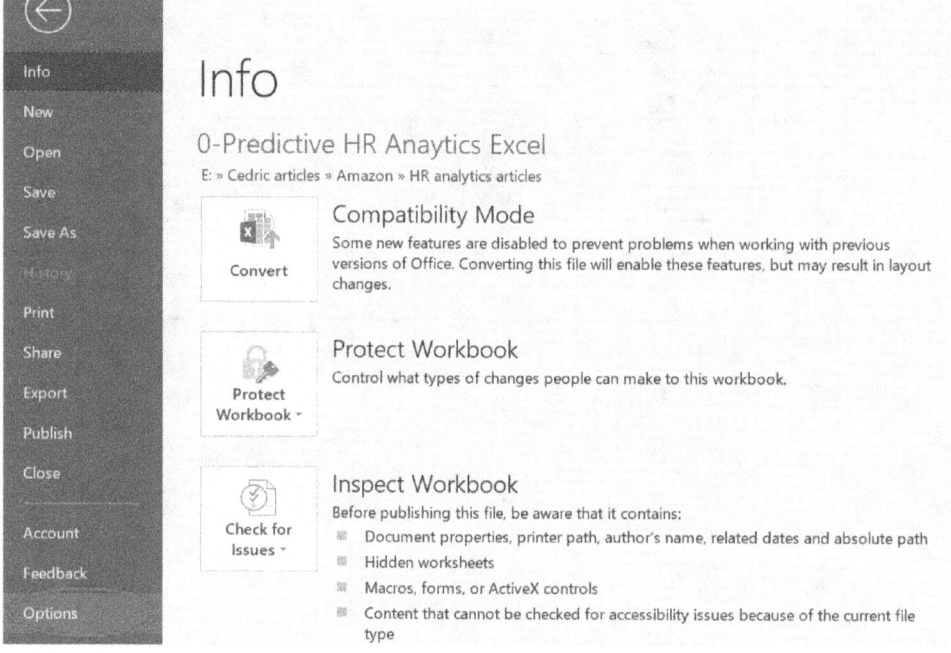

Under Add-ins, click Analysis ToolPak and click the "Go" button.

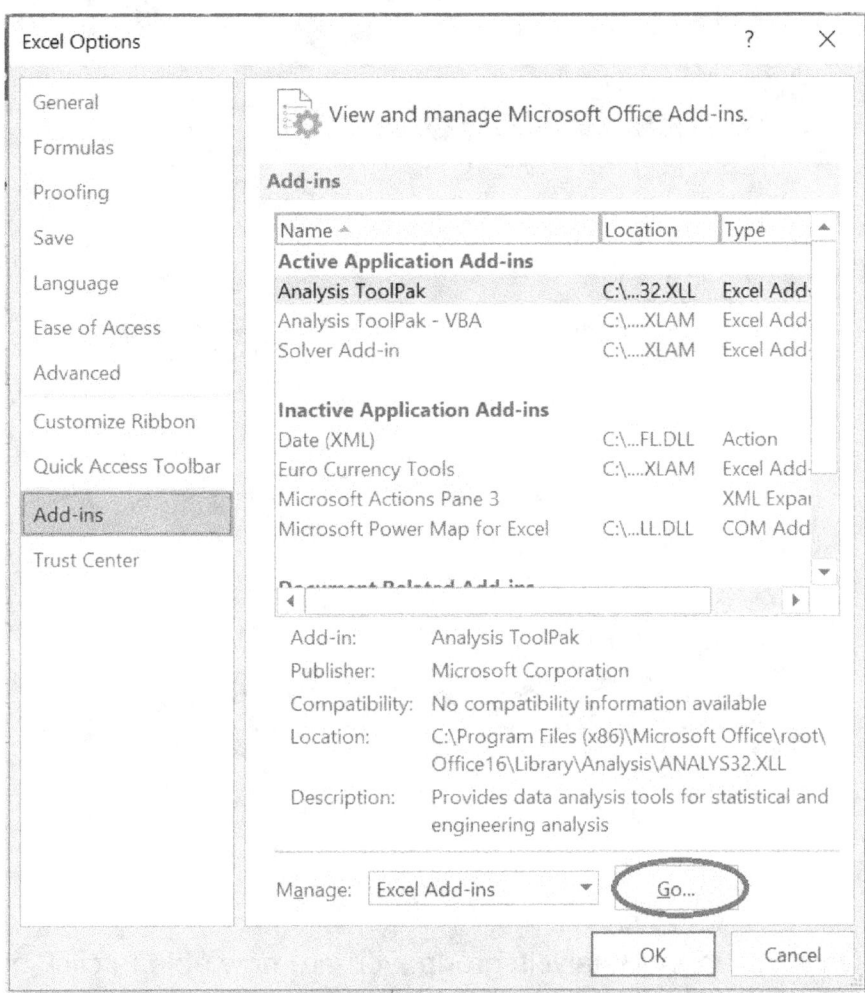

Click "Analysis ToolPak" and click on OK.

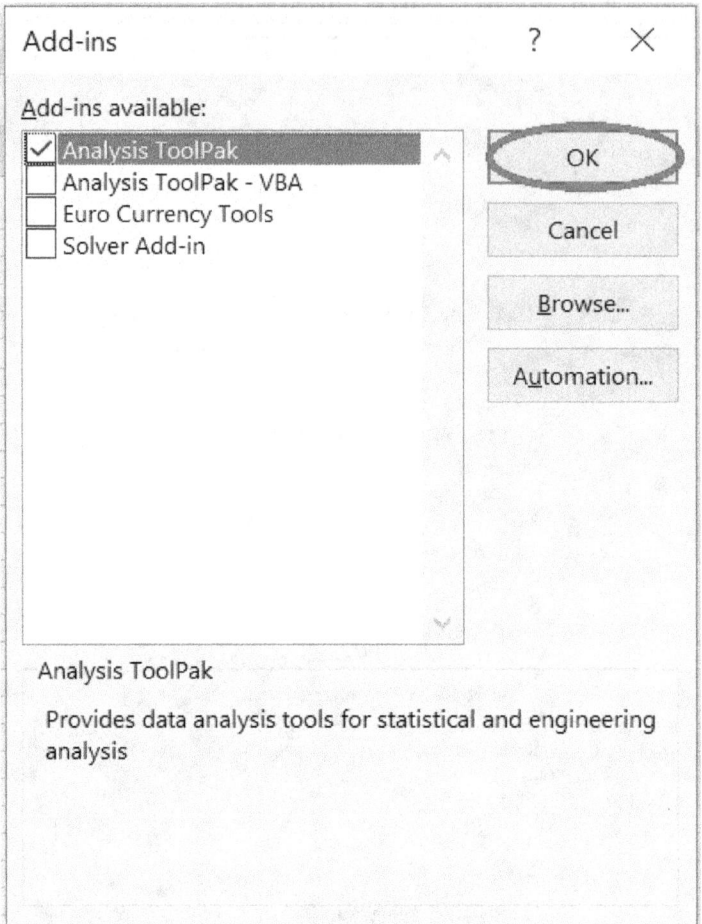

On the Data tab, in the Analysis group, you are now able to click on "Data Analysis".

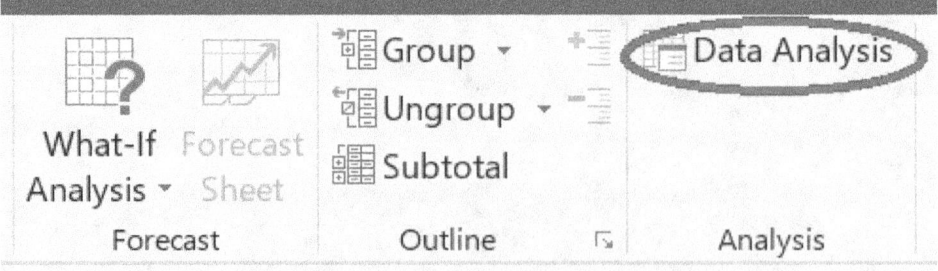

2) Copy the example data in the following table, and paste it in cell A1 of a new Excel worksheet.

	A	B	C	D	E	F	G
1		IISS model's 5 Engagement Fertilizers Score					
2	Year	Basic Needs Score	Social Needs Score	Growth Needs Score	Meaning Score	Expectations Score	Revenue per headcount
3	1	8.3	7.8	9.6	8.8	9.5	1600
4	2	8.5	7.5	8.9	8.3	9.2	1500
5	3	6.8	7.2	9.3	8.2	8.7	1400
6	4	6.5	6.8	9.2	7.9	8.3	1300
7	5	6.2	6.7	9.1	7.4	7.9	1200
8	6	5.9	6.3	8.9	7.3	7.3	1100
9	7	5.7	6.1	8.8	7.1	6.9	1000
10	8	5.5	5.8	8.7	6.8	6.4	900
11	9	5.3	5.6	8.6	6.5	5.8	800
12	10	4.5	5.3	6.2	6.2	5.5	700

3) On the Data tab, in the Analysis group, click on "Data Analysis".

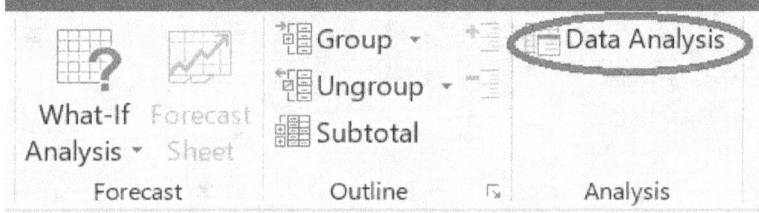

4) Select "Regression" and click "OK".

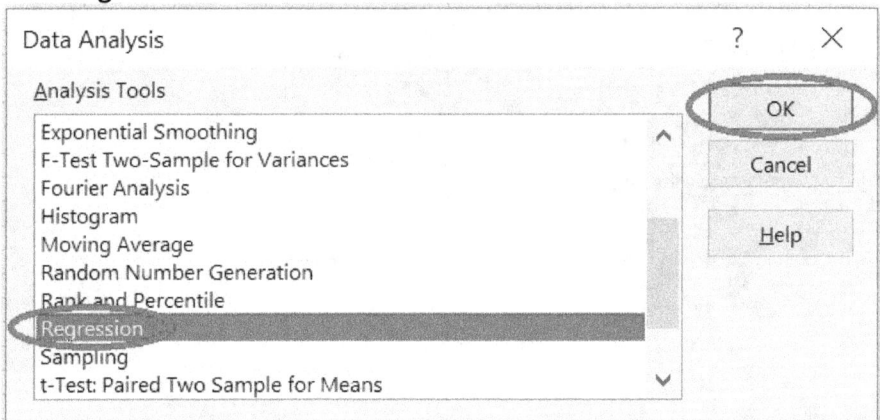

5) After you click OK in the "Data Analysis" dialog box, you will see a "Regression" dialog box.
6) For "Input Y Range", select cells (G2:G12). This is the predictor variable or dependent variable.
7) For "Input X Range", select cells (B2:F12). These are the explanatory variables or independent variables.
8) Check "Labels" box.
9) Click the "Output Range" box, and select cell A14.
10) Click "OK".

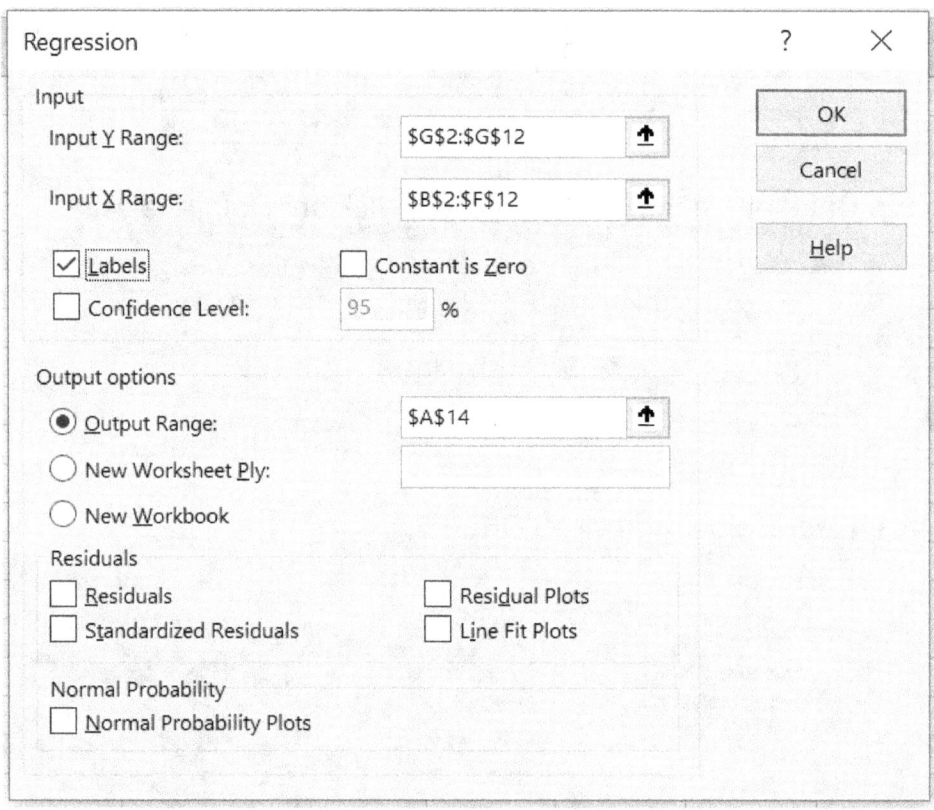

After you click "OK", Excel generates the following Summary Output. Round the numbers to 3 decimal places.

SUMMARY OUTPUT

Regression Statistics	
Multiple R	1.000
R Square	1.000
Adjusted R Square	0.999
Standard Error	8.331
Observations	10

ANOVA

	df	SS	MS	F	ignificance F
Regression	5	824722.393	164944.479	2376.665	0.000
Residual	4	277.607	69.402		
Total	9	825000			

	Coefficients	Standard Error	t Stat	P-value	Lower 95%	Upper 95%	Lower 95.0%	Upper 95.0%
Intercept	-906.599	94.625	-9.581	0.001	-1169.319	-643.879	-1169.319	-643.879
Basic Needs Score	6.626	8.254	0.803	0.467	-16.292	29.544	-16.292	29.544
Social Needs Score	81.727	44.939	1.819	0.143	-43.044	206.498	-43.044	206.498
Growth Needs Score	2.848	4.396	0.648	0.552	-9.358	15.055	-9.358	15.055
Meaning Score	82.570	25.571	3.229	0.032	11.575	153.566	11.575	153.566
Expectations Score	111.611	21.771	5.127	0.007	51.166	172.056	51.166	172.056

R Square: In the output, R Square is 1.000, which means it is a good fit. 100% of the variation in Revenue per headcount (Output) is explained by the independent variables (Input), Basic Needs Score, Social Needs Score, Growth Needs Score, Meaning Score, and Expectations Score. The closer R Square is to "1", the better the regression line fits the data.

Significance F and P-values: To determine if your results are statistically significant (i.e. reliable), check "Significance F" (0.001). If the value of "Significance F" is less than 0.05, it is statistically significant (i.e. reliable). If "Significance F" is bigger than 0.05, don't use this set of independent variables. Delete those variables with "P-value" that is bigger than 0.05 and run the regression again until "Significance F" drops below 0.05. Most or all your P-values should be lower than 0.05. In our example below "P-value" is -906.599, 6.626, 81.727, 2.848, 82.570, and 111.611 for Intercept, Basic Needs Score, Social Needs Score, Growth Needs Score, Meaning Score, and Expectations Score respectively.

Coefficients

From the Summary Output, the regression line is:

SUMMARY OUTPUT

Regression Statistics

Multiple R	1.000
R Square	1.000
Adjusted R Square	0.999
Standard Error	8.331
Observations	10

ANOVA

	df	SS	MS	F	Significance F
Regression	5	824722.393	164944.479	2376.665	0.000
Residual	4	277.607	69.402		
Total	9	825000			

	Coefficients	Standard Error	t Stat	P-value	Lower 95%	Upper 95%	Lower 95.0%	Upper 95.0%
Intercept	-906.599	94.625	-9.581	0.001	-1169.319	-643.879	-1169.319	-643.879
Basic Needs Score	6.626	8.254	0.803	0.467	-16.292	29.544	-16.292	29.544
Social Needs Score	81.727	44.939	1.819	0.143	-43.044	206.498	-43.044	206.498
Growth Needs Score	2.848	4.396	0.648	0.552	-9.358	15.055	-9.358	15.055
Meaning Score	82.570	25.571	3.229	0.032	11.575	153.566	11.575	153.566
Expectations Score	111.611	21.771	5.127	0.007	51.166	172.056	51.166	172.056

Revenue per headcount (Output)
*= - 906.599 + 6.626 * Basic Needs Score + 81.727 * Social Needs Score + 2.848 * Growth Needs Score + 82.570 * Meaning Score + 111.611 * Expectations Score*

Based on the above regression formula,
- For each unit increase in Basic Needs Score, Revenue per headcount increase by 6.626.
- For each unit increase in Social Needs Score, Revenue per headcount increase by 81.727.
- For each unit increase in Growth Needs Score, Revenue per headcount increase by 2.848.
- For each unit increase in Meaning Score, Revenue per headcount increase by 82.570.
- For each unit increase in Expectations Score, Revenue per headcount increase by 111.611.

Coefficients can also be used for forecasting. For example, if "*Basic Needs Score*", "*Social Needs Score*", "*Growth Needs Score*", "*Meaning Score*" and "*Expectations Score*" are all 7.0, then **predicted Revenue per headcount (Output)**
= - 906.599 + 6.626 * *Basic Needs Score* + 81.727 * *Social Needs Score* + 2.848 * *Growth Needs Score* + 82.570 * *Meaning Score* + 111.611 * *Expectations Score*
= - 906.599 + 6.626*7 + 81.727*7 + 2.848*7 + 82.570*7 + 111.611*7
= **$1091**

5.7 Sentiment analysis

Sentiment Analysis (also known as opinion mining) is the process of mathematically categorizing opinions expressed in text, to determine whether the attitude towards a company, product, or topic, is positive, negative, or neutral. Text information is constantly growing in review sites, forums, and social media. By using sentiment analysis, you gauge how people feel about your company without having to read through thousands of comments.

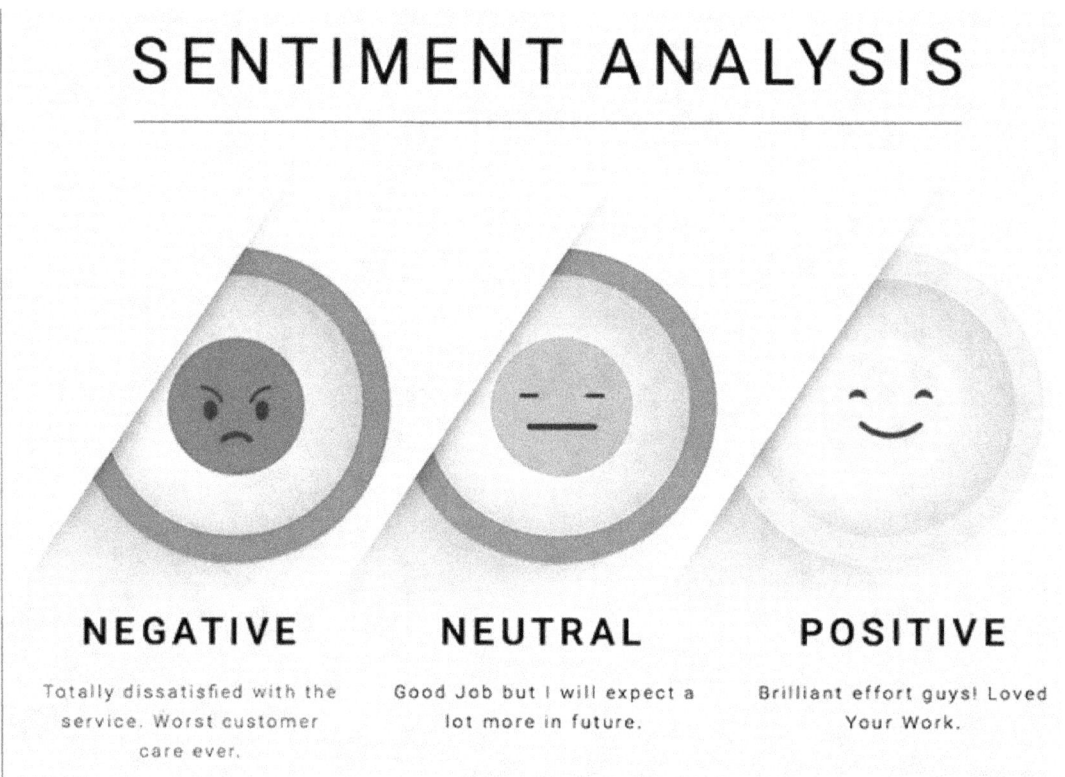

Source: *https://www.kdnuggets.com/2018/03/5-things-sentiment-analysis-classification.html*

People's engagement with business and brand perception depends heavily on public opinion. According to a survey by Podium, 93 percent of consumers say that online reviews influence their buying decisions. Sentiment analysis allows companies to monitor and measure people's attitude towards an organization so that they can address it timely. [1]

Customer Feedback text	Sentiment
This restaurant is excellent! The waiters are very friendly and the steak is delicious!	Positive
I will not recommend this restaurant to anyone. The steak is horrible and expensive, and the service is poor.	Negative

How Sentiment Analysis works

Rules-based sentiment analysis technique and uses a dictionary of words labelled by sentiment to determine the sentiment (-1 = Negative, +1 = Positive) of a sentence: [2]

- **Define two lists of polarized words.**
  E.g. Negative words such as bad, worst, ugly, etc. Positive words such as excellent, best, beautiful, etc.
- **Count the number of positive and negative words.**
  Given a text, count the number of positive words, and negative words. If the number of positive word appearances is greater than the number of negative word appearances return a positive sentiment, conversely, return a negative sentiment. Otherwise, return neutral.

Sentence	Sentiment	Score
great colleagues	positive	0.82
nasty bosses, long hours	negative	0.42

Sentiment Analysis limitations

Sentiment Analysis has a few limitations:

- **Context and Polarity**
  Analyzing sentiment without context is difficult because of changes in polarity. If the responses to a survey question, "What did you like about the event?", is "Everything of it" and "Absolutely nothing!", the first response would be positive and the second one would be negative. But, if the responses come from answers to the question, "What did you Dislike about the event?", the negative in the question will make sentiment analysis change altogether! [2]

- **Irony and Sarcasm**
  If the response to a survey question, "Have you had a nice customer experience with us?", is "Yeah, sure.", this likely to be classified as negative because "yeah and sure "belong to positive or neutral texts. But, in reality it might be a negative sentiment from a customer using irony and sarcasm. [2]

- **Comparisons**
  How to treat comparisons such as "This is better than old tools" in sentiment analysis is another challenge, where context makes a difference. Would you classify them as neutral or positive? At first glance, it seems to be a positive sentiment. But if the old tools are useless, then it is a neutral sentiment. [2]

Sentiment Word Cloud

To visualize the results of Sentiment Analysis, you can use graphs, histograms, and Word Cloud.

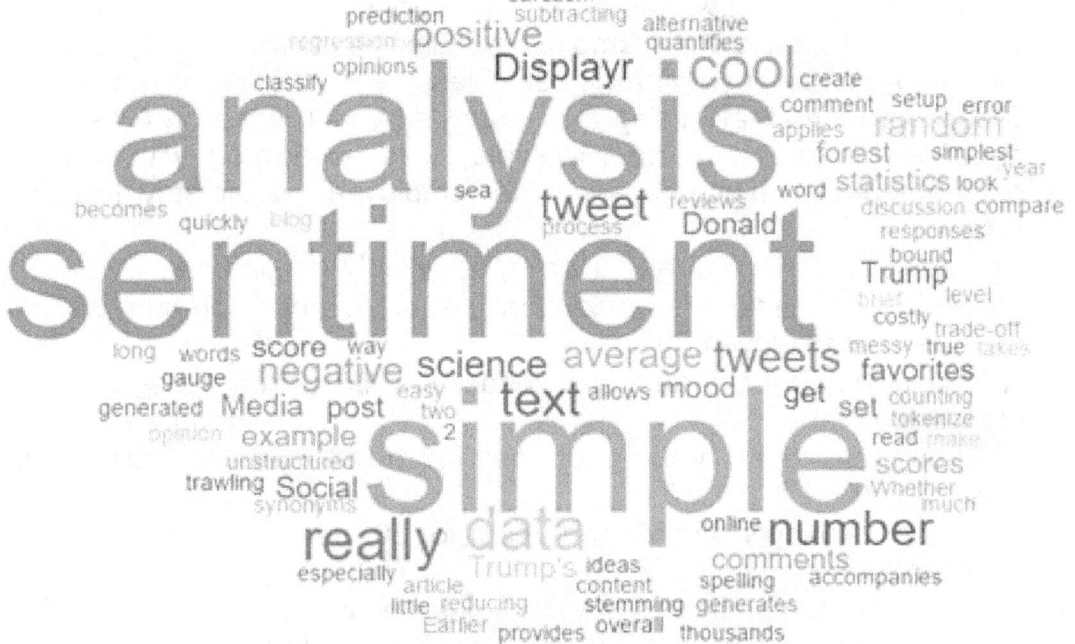

Source: https://www.kdnuggets.com/2018/03/5-things-sentiment-analysis-classification.html

Reference
(1) 2017 State of Online Reviews (2017) Consumers Get "Buy" with a little help from their friends. http://learn.podium.com/rs/841-BRM-380/images/2017-SOOR-Infographic.jpg (13 June 2019)
(2)   Monkeylearn (2019) Sentiment Analysis - Nearly Everything You Need to Know. https://monkeylearn.com/sentiment-analysis/ (13 June 2019)

5.7.1 Real-World Impact of Sentiment Analysis

Research have shown that Employer Branding and Sentiment Analysis has an impact on employee salaries, number of job applicants, cost per hire, retention, and employee turnover:

- Glassdoor found that workplace culture matters for employee retention. When employees switch employers, Glassdoor found they usually move to companies with higher Glassdoor ratings. In particular, Glassdoor found that raising a company's overall rating on Glassdoor by one star (on a one-to-five scale) was associated with a four-percentage-point higher chance that employees would stay for their next role. [1]
- Companies with weak brand overpay salaries by 10 percent. [2]
- 69 percent of candidates wouldn't take a job at a bad company even if they were unemployed. [2]
- A strong employer brand can lead to a 50 percent decrease in cost/hire. [2]
- A strong employer brand can lead to a 28 percent increase in retention. [2]
- A strong employer brand can lead to 50 percent more qualified applicants. [2]
- 92 percent of candidates would consider leaving their jobs if a company with an excellent corporate reputation offered them. [2]
- Companies with a strong talent brand get up to 2.5 times more applicants per job post on LinkedIn. [2]

Reference
(1) Dr. Andrew Chamberlain (2017) Why Do Employees Stay? A Clear Career Path and Good Pay, for Starters. https://www.glassdoor.com/research/why-do-employees-stay-a-clear-career-path-and-good-pay-for-starters/ (26 November 2018)
(2) getfive (2018) A Cautionary Tale About Bad Glassdoor Ratings. https://getfive.com/blog/a-cautionary-tale-about-bad-glassdoor-ratings/ (26 November 2018)

5.7.2 Run Sentiment Analysis in Excel with Azure Machine Learning

This section teaches you how to run sentiment analysis with Azure Machine Learning.

It would be tedious if you have to go through hundreds of survey comments to assess employee sentiment. There is a free add-in from Microsoft that helps you to do sentiment analysis in Excel, and it can compute a probability showing how positive or negative each comment is. It has a dictionary of 5,097 negative and 2,533 positive words, and each word is assigned a strong or weak polarity. [1]

1) Copy the text below into cells A1 to A9 of your Excel spreadsheet:

	A
1	**Glassdoor review**
2	pay is ok, friendly colleagues
3	nice colleagues
4	good learning ground for fresh graduates
5	good worklife
6	not much to learn
7	slow progression, low pay and a lot of work
8	entry level salary is not competitive
9	bad management, politics and work culture

2) Go to "Insert" tab, click "Get Add-ins":

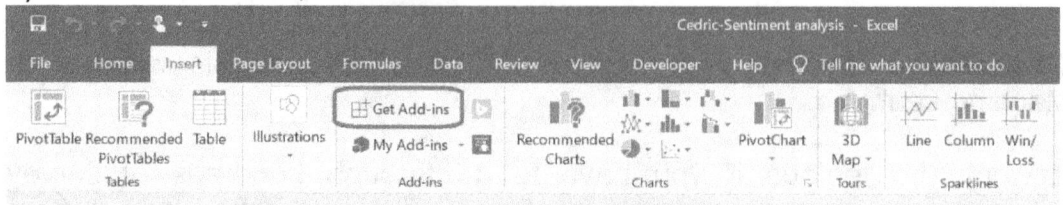

3) Search for "Azure Machine", click add:

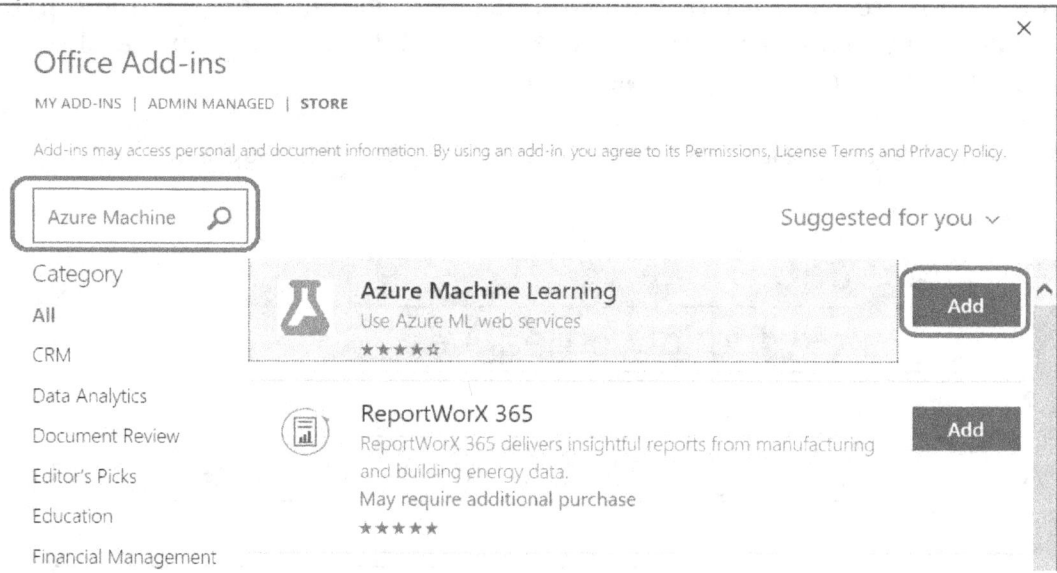

4) Click "Text Sentiment Analysis":

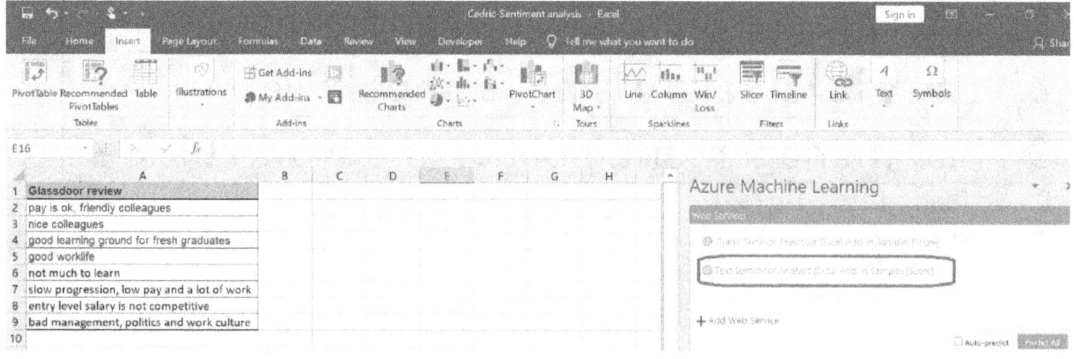

5) Click "View Schema":

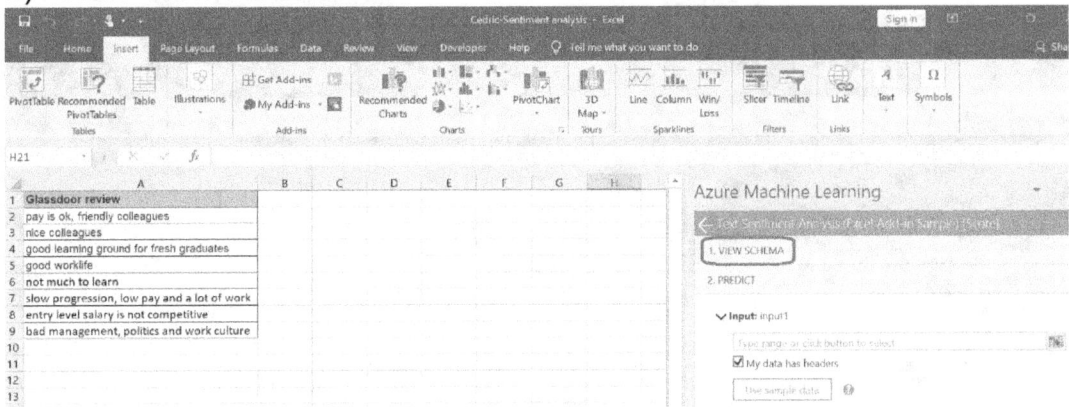

6) Replace your "Glassdoor review" heading word with the Schema word "tweet_text":

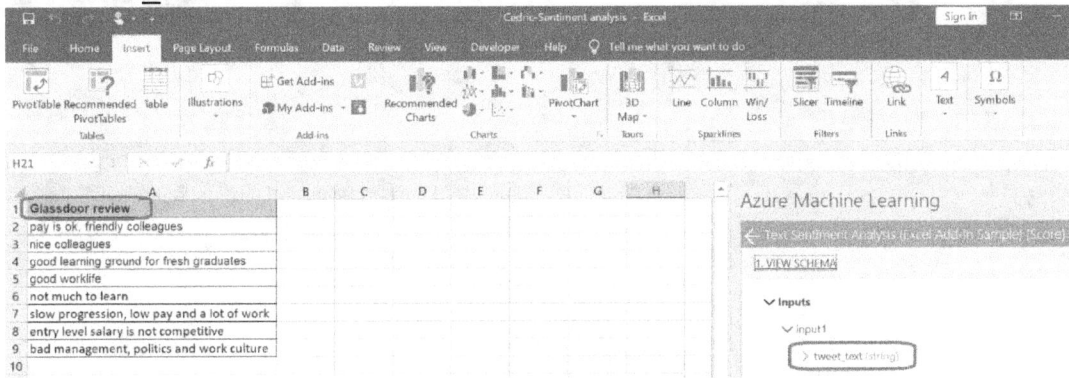

7) Replace the header "Glassdoor review" with the word "tweet_text". Take note that the word "tweet_text" has to be case sensitive.

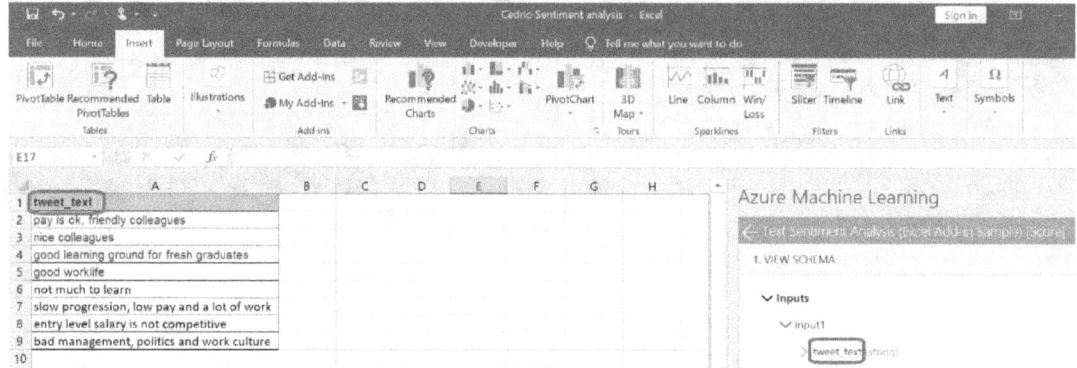

8) Click "View Scema" to close it.

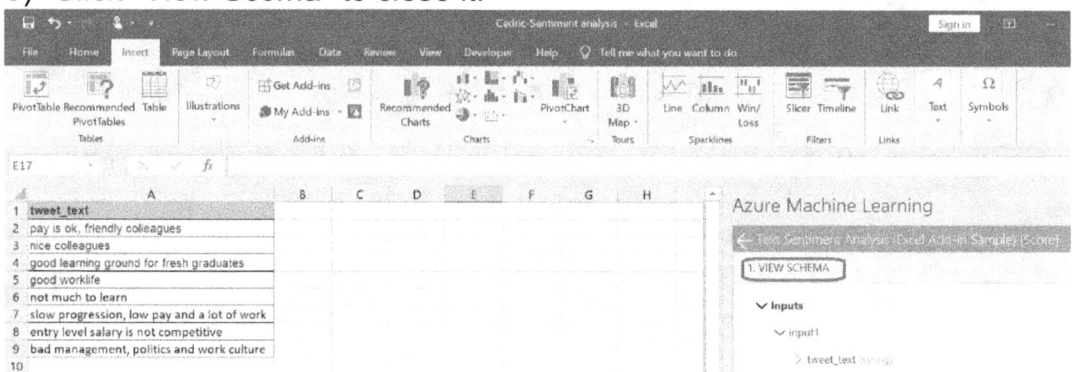

9) Select cells "A1:A9" and tick "My data has headers" for "Input". Input "B1" and tick "Include headers" for "Output". Make sure you have 2 blank columns beside column A. Then click "Predict".

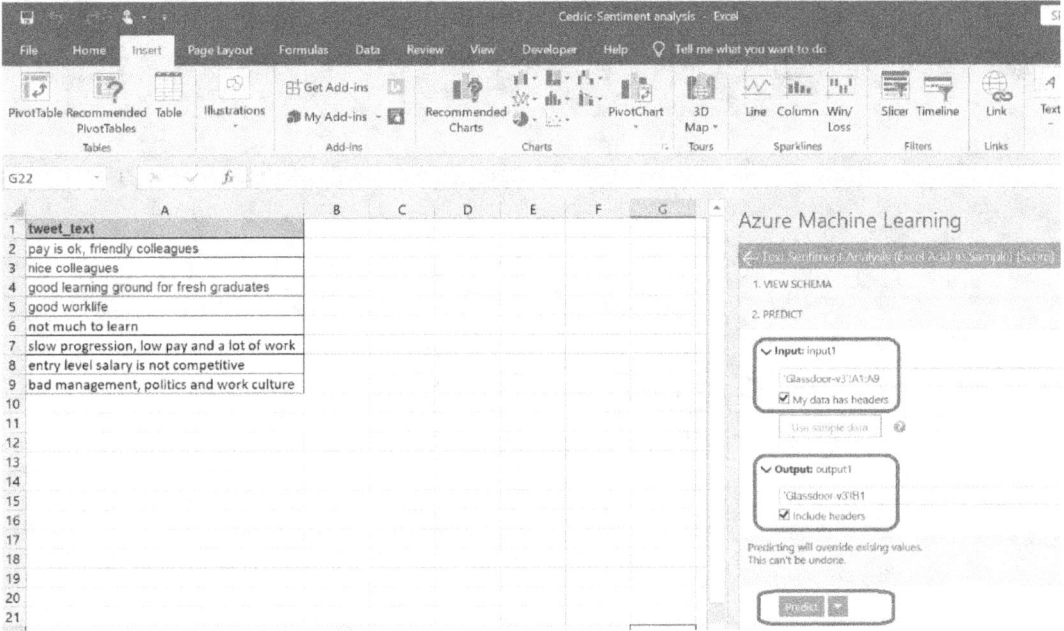

10) After you click "Predict", the "Sentiment" and "Score" column appears. Select Column C, the click "%" at the home bar to convert "Score" to percentages.

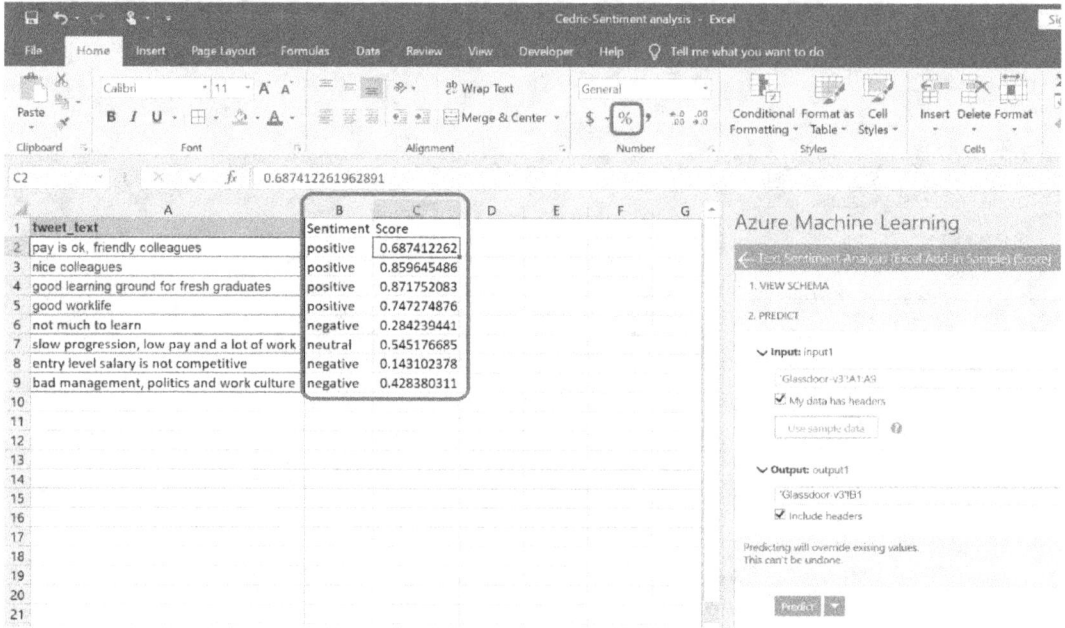

11) Sentiment ranges from 0% to 100%. 100% means very positive sentiment. 0% means very negative sentiment.

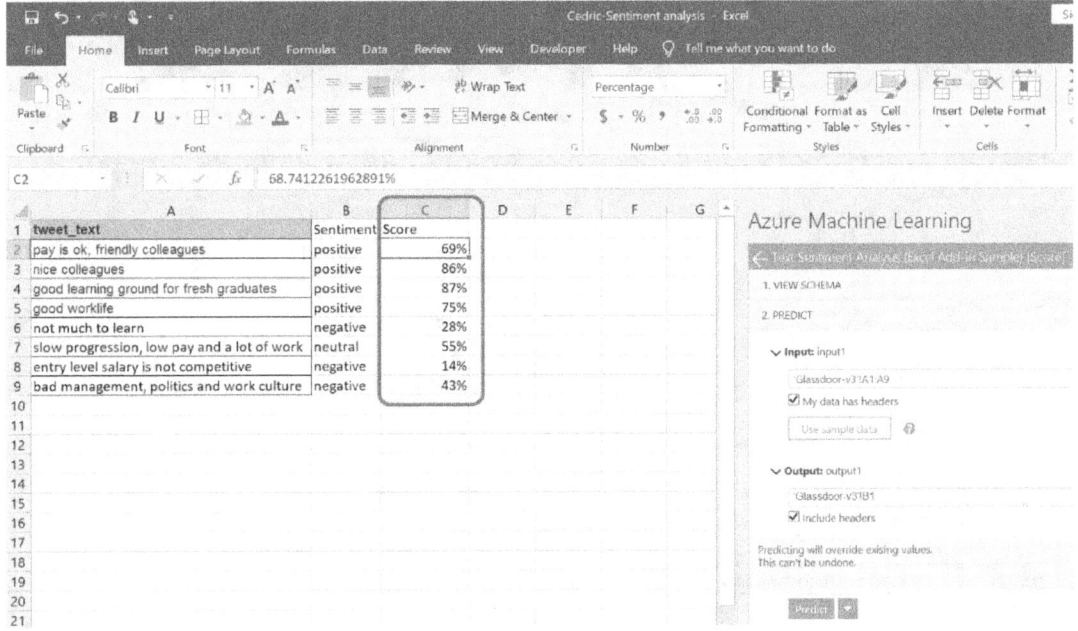

12) Click cell C2, then at "data tab" select sort "Z to A" to sort the "Score from highest to lowest.

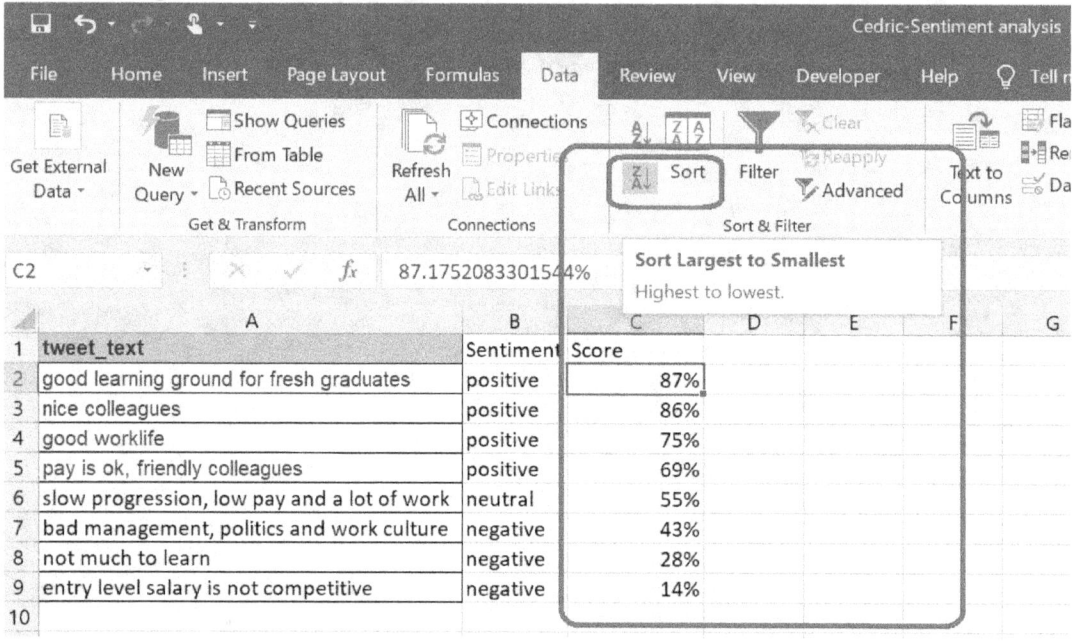

13) To Insert pivot table, to "Insert" tab, click "PivotTable". Under "Select a table or range", select cells A1 to C9. Under "Existing worksheet" enter cell E2 for "location", and click "OK".

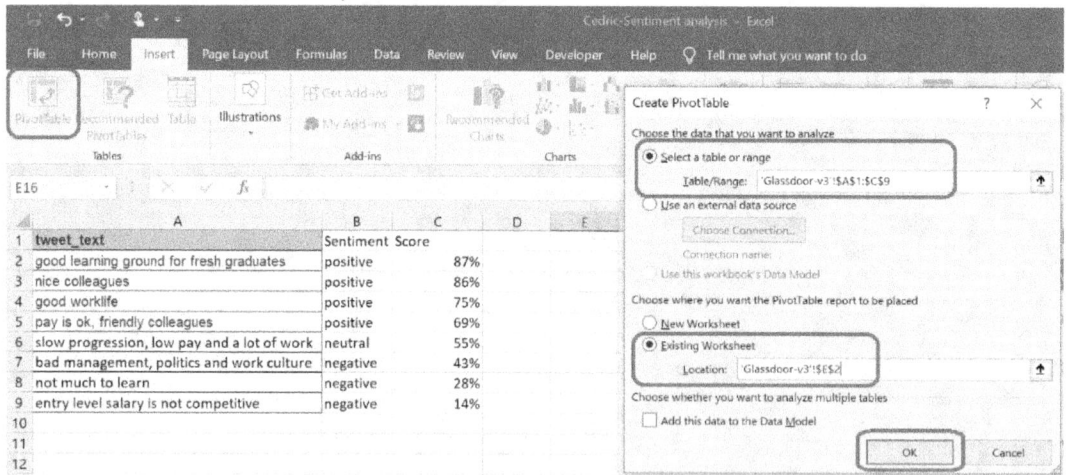

14) At the "PivotTable Fields", click "Sentiment" and "Score".

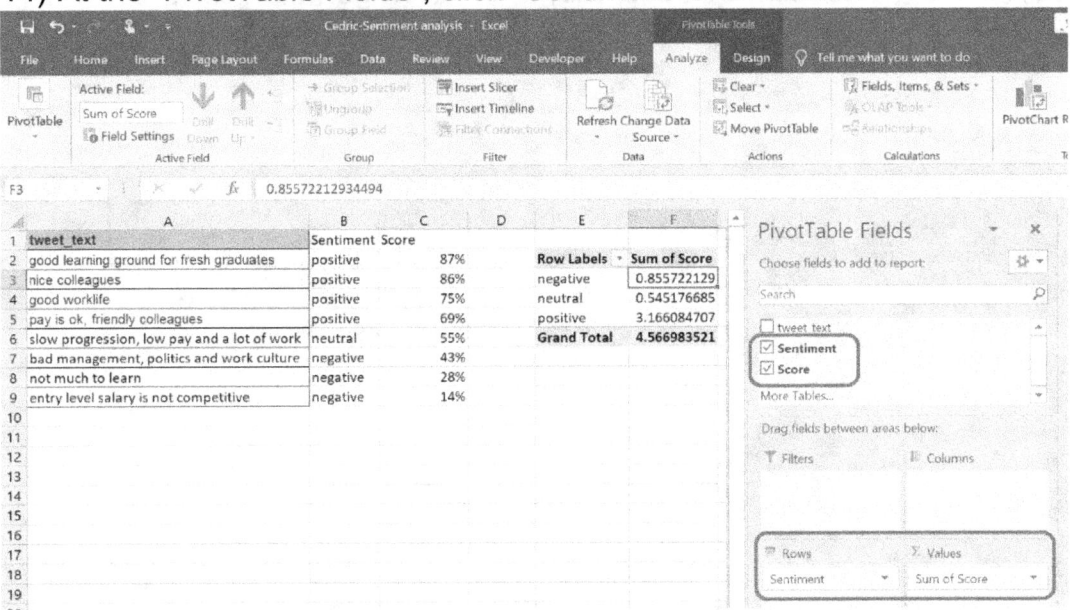

15) Click "Sum of Score", then click "Value Field Settings".

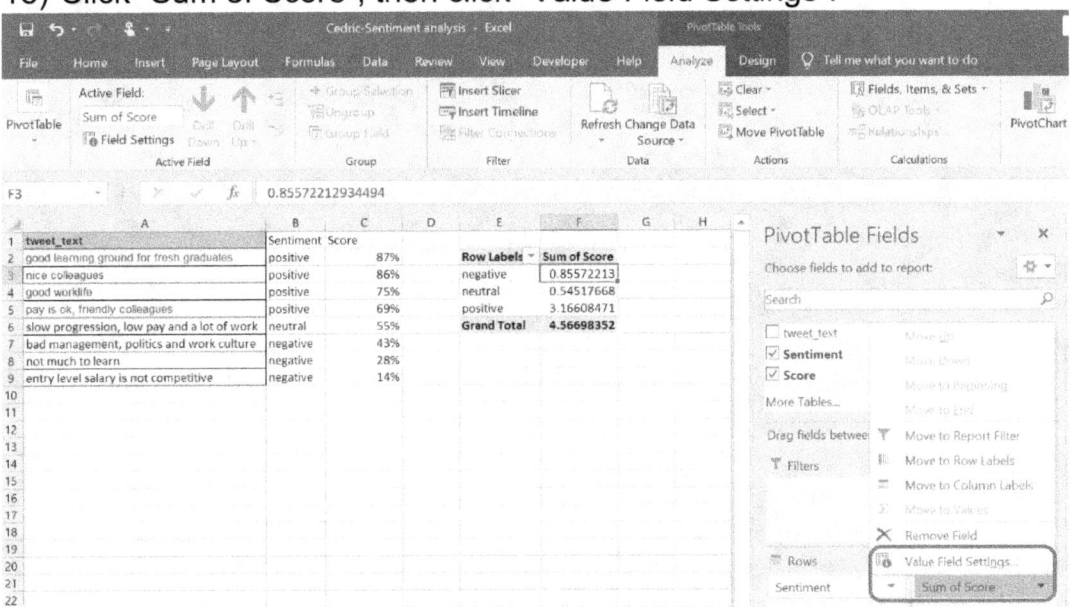

16) At the "Value Field Settings" tab, click "Average", then click "OK".

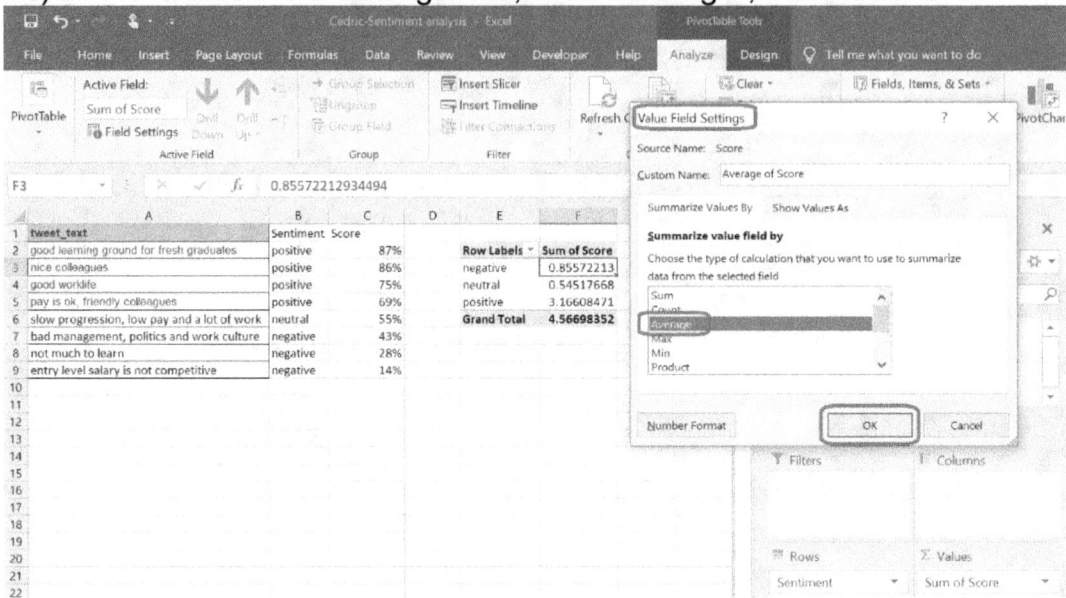

17) Check "tweet_text" and move it to "Values". You'll see "Count of tweet_text".

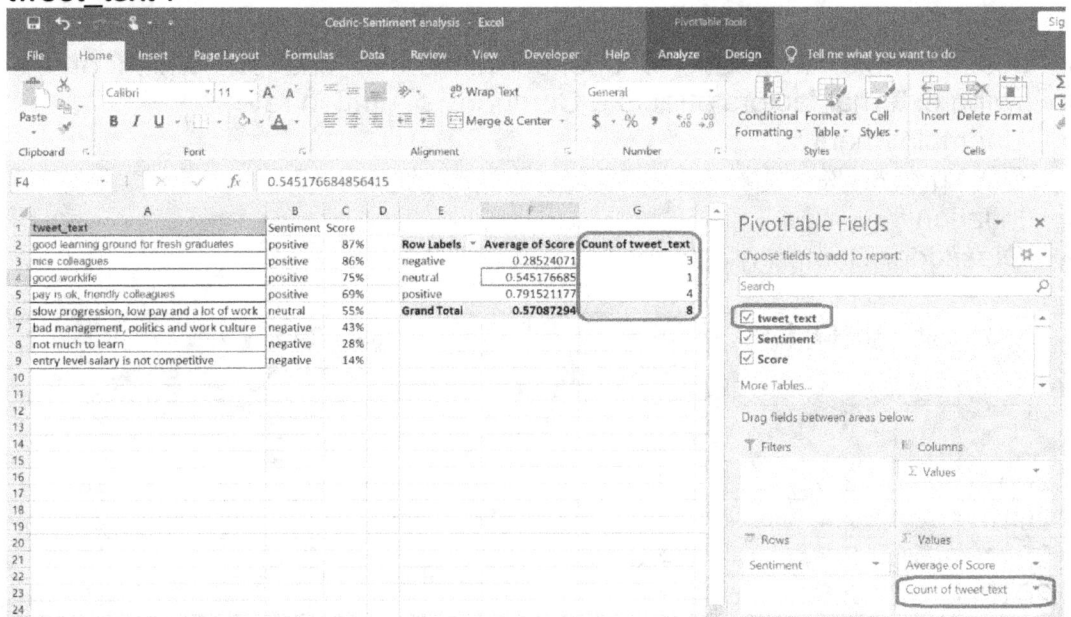

References:
(1) Bill Jelen (2017), Sentiment Analysis. https://www.mrexcel.com/excel-tips/sentiment-analysis/ (3 June 2019)

5.7.3 Correlation Example: Determine relationship between "Glassdoor Company Ratings" and "Company Attrition Rate".

In this example, we want to use correlation to find out which of the following external factors affects a Company's Attrition rate:.
- Unemployment %
- GDP growth %
- Inflation %
- Glassdoor Company ratings

Year	Unemployment %	GDP growth %	Inflation %	Company ABC's Glassdoor rating	Company ABC's attrition rate
2009	2.3	4.1	2.3	1	20
2010	2.4	3.9	2.3	1	21
2011	2.3	4.3	2.3	2	18
2012	2.5	3.5	2.3	2	15
2013	2.6	2.8	2.2	2	17
2014	2.8	2.4	2.3	3	13
2015	3.1	2.6	2.2	3	13
2016	2.9	2.5	2.3	3	12
2017	3.2	2.1	2.2	4	12
2018	3.3	2.3	2.3	4	11
2019	3.4	2.2	2.3	4	11

1) Install "Analysis ToolPak", an Excel add-in

"Analysis ToolPak" is an add-in for Microsoft Excel that comes with Microsoft Excel. To be able to run regression using Excel, you need to first install "Analysis ToolPak", an Excel add-in program that provides data analysis tools. To load the Analysis ToolPak add-in, follow these steps:

- On the File tab, click Options.

- Under Add-ins, click Analysis ToolPak and click the "Go" button.

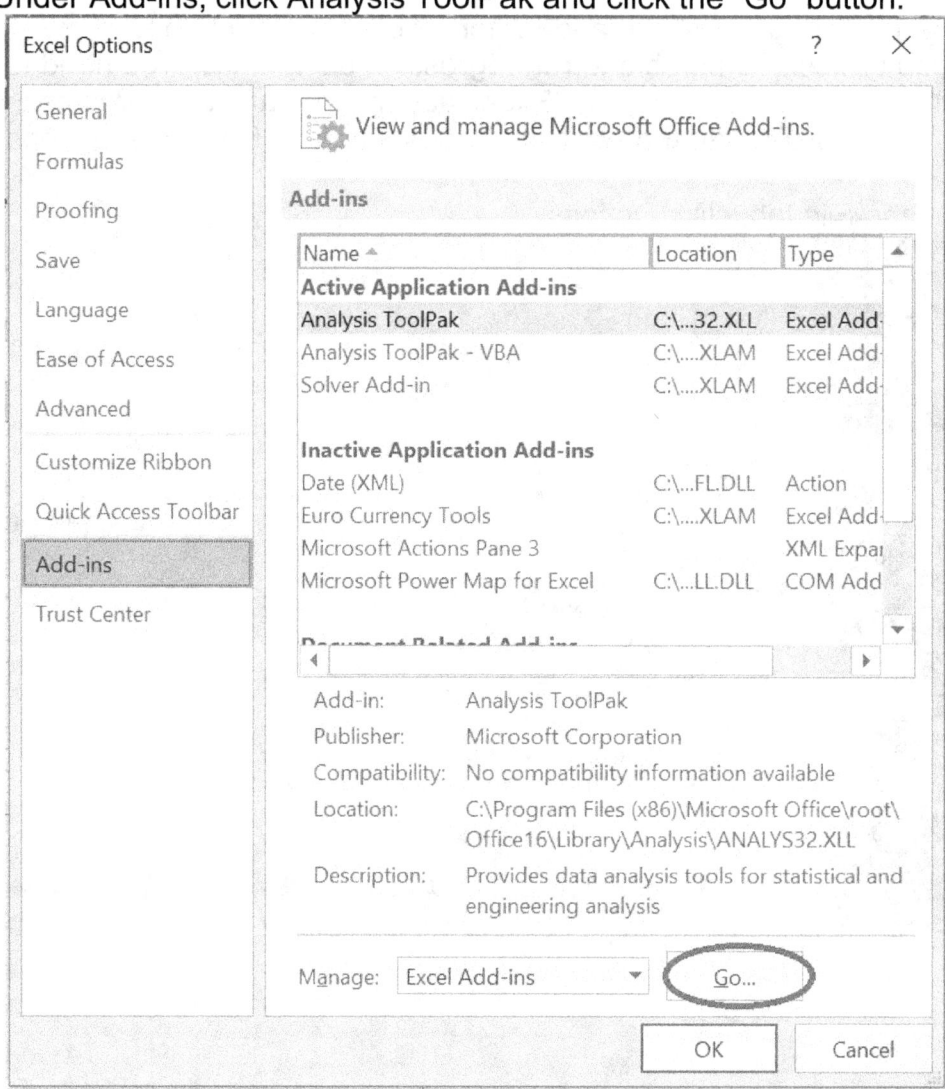

- Click "Analysis ToolPak" and click on OK.

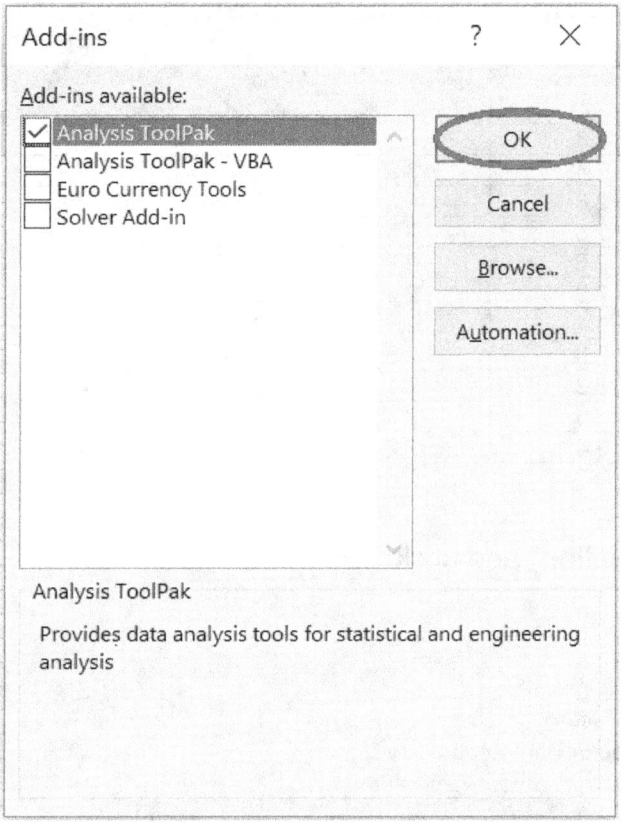

- On the Data tab, in the Analysis group, you are now able to click on "Data Analysis".

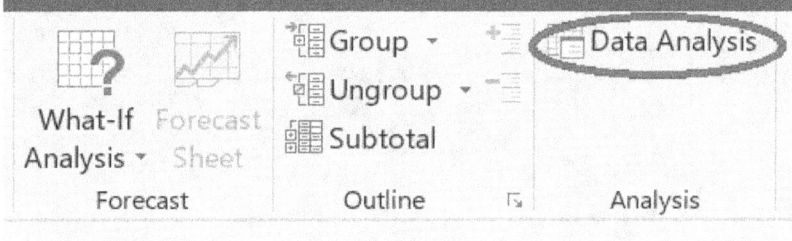

2) Copy the example data in the following table, and paste it in cell A1 of a new Excel worksheet.

	A	B	C	D	E	F
1	Year	Unemployment %	GDP growth %	Inflation %	Company ABC's Glassdoor rating	Company ABC's attrition rate
2	2009	2.3	4.1	2.3	1	20
3	2010	2.4	3.9	2.3	1	21
4	2011	2.3	4.3	2.3	2	18
5	2012	2.5	3.5	2.3	2	15
6	2013	2.6	2.8	2.2	2	17
7	2014	2.8	2.4	2.3	3	13
8	2015	3.1	2.6	2.2	3	13
9	2016	2.9	2.5	2.3	3	12
10	2017	3.2	2.1	2.2	4	12
11	2018	3.3	2.3	2.3	4	11
12	2019	3.4	2.2	2.3	4	11

3) Select "Correlation" and click "OK".

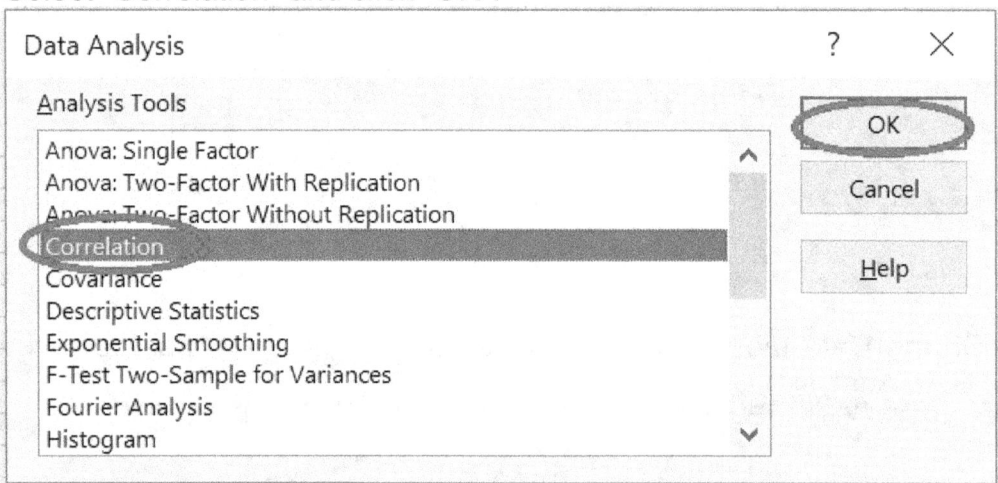

4) After you click OK in the "Data Analysis" dialog box, you will see a "Correlation" dialog box.
5) For "Input Range", select cells (B1:F12).
6) Check "Labels in first row".
7) For "Output Range", select cells (A14).
8) Click "OK".

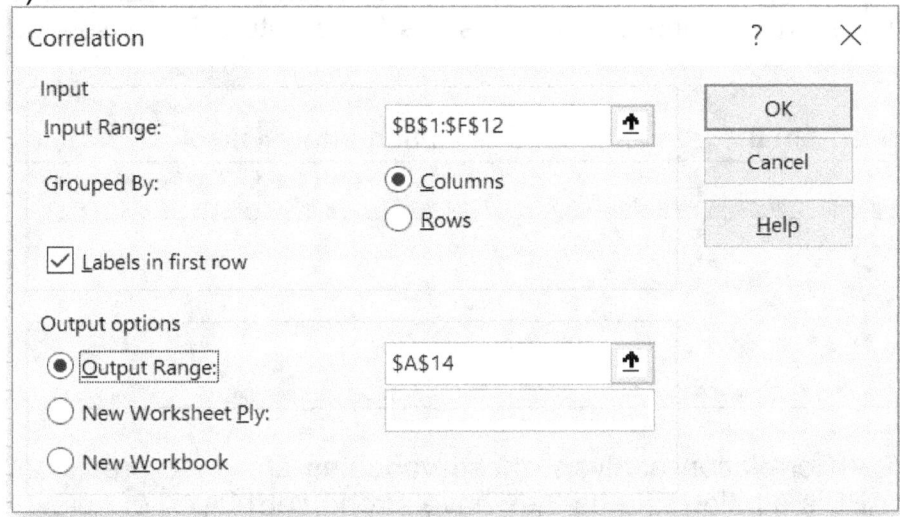

After you click "OK", Excel generates the following Correlation analysis.

	Unemployment %	GDP growth %	Inflation %	Company ABC's Glassdoor rating	Company ABC's attrition rate
Unemployment %	1				
GDP growth %	-0.91	1			
Inflation %	-0.26	0.37	1		
Company ABC's Glassdoor rating	0.94	-0.88	-0.21	1	
Company ABC's attrition rate	-0.90	0.89	0.14	-0.95	1

A negative correlation coefficient means that an increase in X is associated with a decrease in Y. Similar to a positive correlation, a negative correlation shows a connection between two variables, and the relative strengths are the same. In other words, a correlation coefficient of 0.85 has the same strength as a correlation coefficient of -0.85. Correlation coefficients are always values between -1 and 1, where "-1" means that there is a perfect linear negative correlation, while "1" shows a perfect linear positive correlation. A correlation coefficient of zero, or near to zero, means that there is no meaningful relationship between variables. Correlation coefficient of 0.91 or -0.92 shows a very strong positive and negative correlation respectively. However, correlation does not mean causation. An example of negative correlation is the amount of snowfall and the temperature. As the temperature increases, the amount of snowfall decreases. An example of positive correlation is the relationship between temperature and ice cream sales. As temperature increases, so do ice cream sales.

9) Observations from the above Excel Correlation analysis:

	Unemployment %	GDP growth %	Inflation %	Company ABC's Glassdoor rating	Company ABC's attrition rate
Unemployment %	1				
GDP growth %	-0.91	1			
Inflation %	-0.26	0.37	1		
Company ABC's Glassdoor rating	0.94	-0.88	-0.21	1	
Company ABC's attrition rate	-0.90	0.89	0.14	-0.95	1

From the Excel Correlation analysis, these variables are good predictors of Company ABC's attrition rate as they have strong correlation of below -0.75 and above 0.75:

- Unemployment %: -0.90 Correlation coefficient with Company ABC's attrition rate.
- GDP growth %: -0.89 Correlation coefficient with Company ABC's attrition rate.
- Company ABC's Glassdoor rating: -0.95 Correlation coefficient with Company ABC's attrition rate.

From the Excel Correlation analysis, Inflation % has very little impact on Company ABC's attrition rate as they have very weak correlation of between -0.20 to 0.20.

5.7.4 Regression Example: Predict "Company Attrition Rate" with "Glassdoor Company Ratings".

In this example, we want to use multiple regression to predict the company's attrition rate, based on changes in external factors (Unemployment %, GDP growth %, Inflation %, and Glassdoor Company ratings).

1) Install "Analysis ToolPak", an Excel add-in

"Analysis ToolPak" is an add-in for Microsoft Excel that comes with Microsoft Excel. To be able to run regression using Excel, you need to first install "Analysis ToolPak", an Excel add-in program that provides data analysis tools. To load the Analysis ToolPak add-in, follow these steps:

On the File tab, click Options.

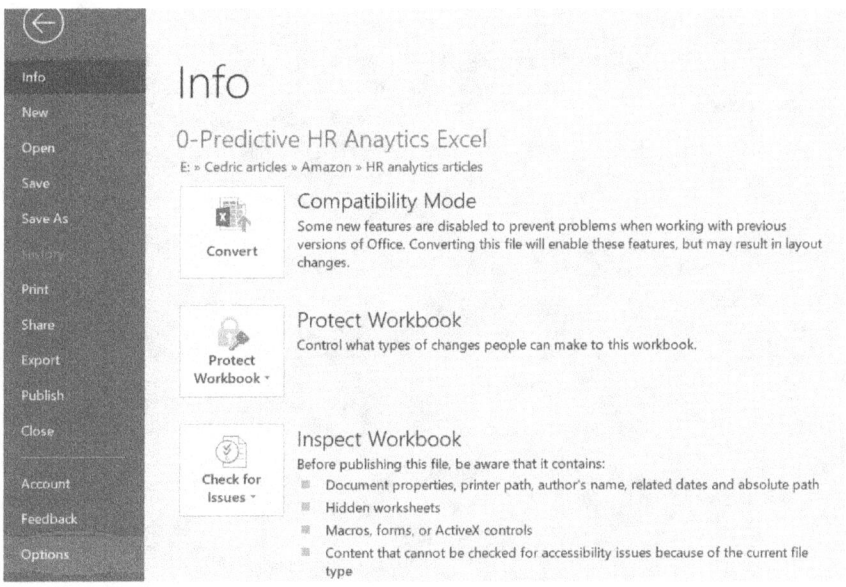

Under Add-ins, click Analysis ToolPak and click the "Go" button.

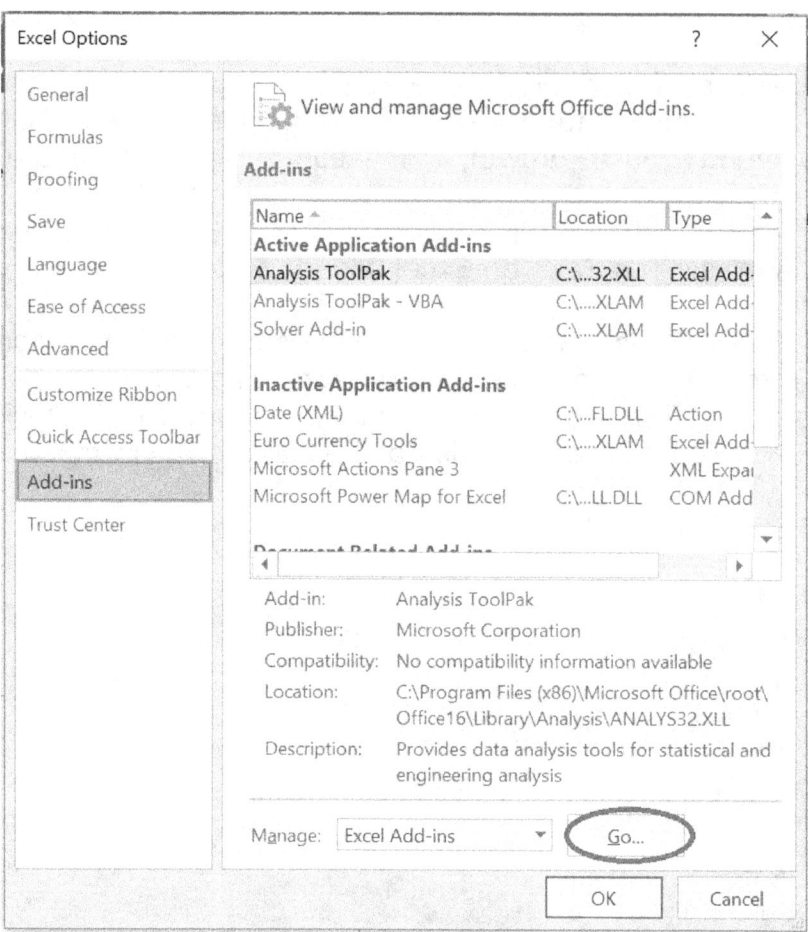

Click "Analysis ToolPak" and click on OK.

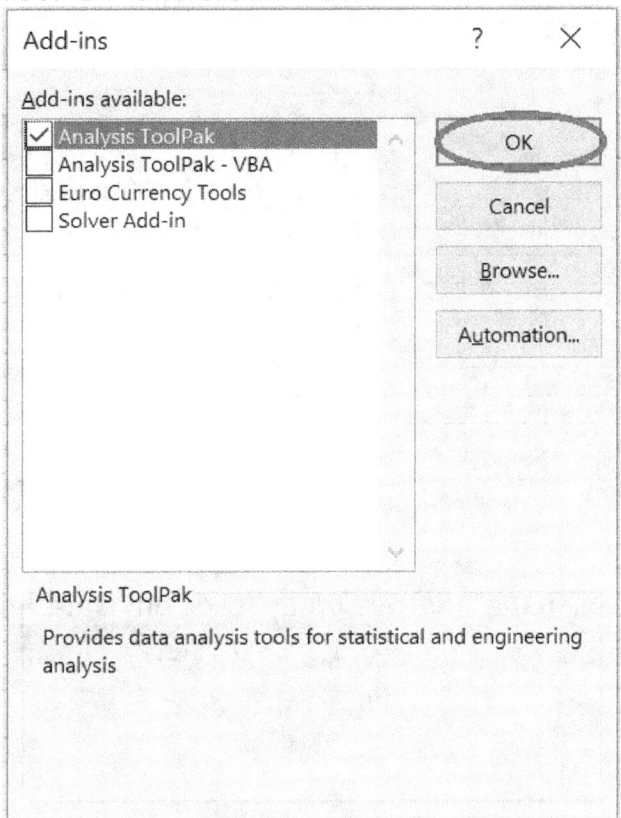

On the Data tab, in the Analysis group, you are now able to click on "Data Analysis".

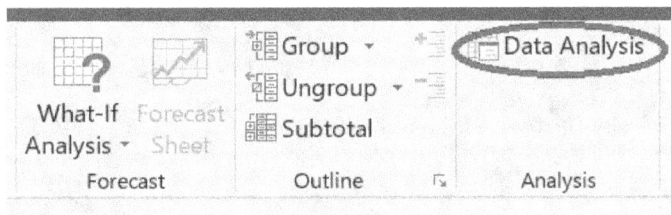

2) Copy the example data in the following table, and paste it in cell A1 of a new Excel worksheet.

	A	B	C	D	E
1	Year	Unemployment %	GDP growth %	Company ABC's Glassdoor rating	Company ABC's attrition rate
2	2009	2.3	4.1	1	20
3	2010	2.4	3.9	1	21
4	2011	2.3	4.3	2	18
5	2012	2.5	3.5	2	15
6	2013	2.6	2.8	2	17
7	2014	2.8	2.4	3	13
8	2015	3.1	2.6	3	13
9	2016	2.9	2.5	3	12
10	2017	3.2	2.1	4	12
11	2018	3.3	2.3	4	11
12	2019	3.4	2.2	4	11

3) On the Data tab, in the Analysis group, click on "Data Analysis".

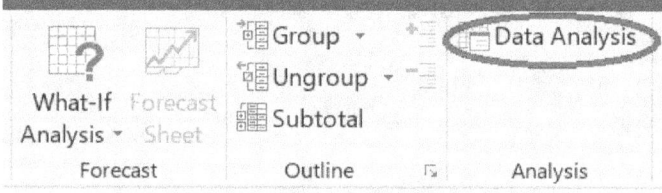

4) Select "Regression" and click "OK".

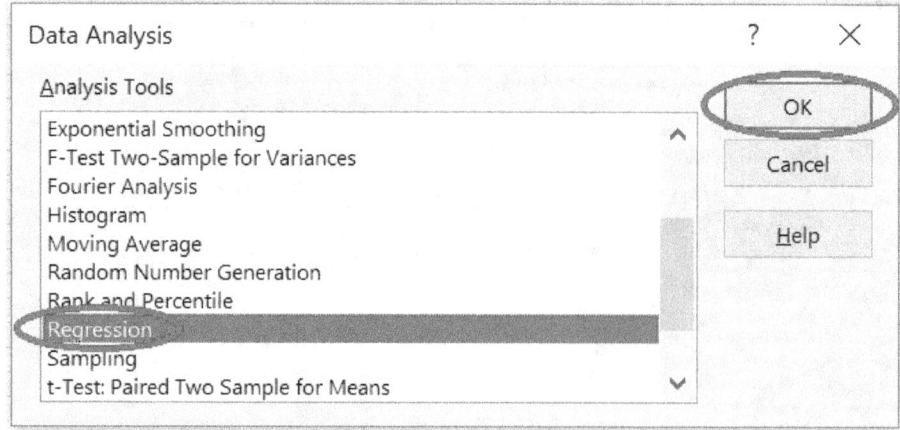

5) After you click OK in the "Data Analysis" dialog box, you will see a "Regression" dialog box.
6) For "Input Y Range", select cells (E1:E12). This is the predictor variable or dependent variable.
7) For "Input X Range", select cells (B1:D12). These are the explanatory variables or independent variables.
8) Check "Labels" box.
9) Click the "Output Range" box and select cell A14.
10) Click "OK".

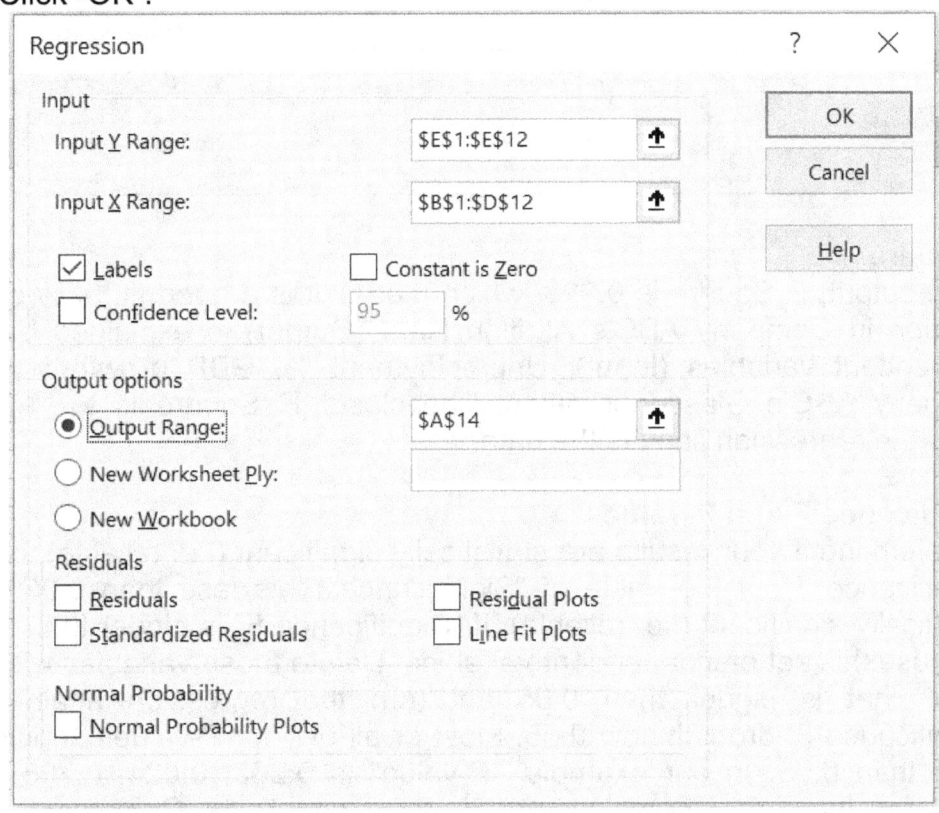

After you click "OK", Excel generates the following Summary Output. Round the numbers to 3 decimal places.

SUMMARY OUTPUT

Regression Statistics	
Multiple R	0.961
R Square	0.923
Adjusted R Square	0.890
Standard Error	1.203
Observations	11

ANOVA

	df	SS	MS	F	Significance F
Regression	3	121.498	40.499	27.963	0.000
Residual	7	10.138	1.448		
Total	10	131.636			

	Coefficients	Standard Error	t Stat	P-value	Lower 95%	Upper 95%	Lower 95.0%	Upper 95.0%
Intercept	13.708	9.814	1.397	0.205	-9.499	36.915	-9.499	36.915
Unemployment %	1.631	3.181	0.513	0.624	-5.890	9.152	-5.890	9.152
GDP growth %	1.316	1.136	1.159	0.285	-1.369	4.001	-1.369	4.001
Company ABC's Glassdoor rating	-2.795	1.036	-2.698	0.031	-5.246	-0.345	-5.246	-0.345

R Square
In the output, R Square is 0.923, which means it is a good fit. 92% of the variation in Company ABC's Attrition Rate (Output) is explained by the independent variables (Input): Unemployment %, GDP growth %, and Company ABC's Glassdoor rating. The closer R Square is to "1", the better the regression line fits the data.

Significance F and P-values
To determine if your results are statistically significant (i.e. reliable), check "Significance F". If the value of "Significance F" is less than 0.05, it is statistically significant (i.e. reliable). If "Significance F" is bigger than 0.05, don't use this set of independent variables. Delete those variables with "P-value" that is bigger than 0.05 and run the regression again until "Significance F" drops below 0.05. Most or all of your P-values should be lower than 0.05. In our example, "P-value" is 0.205, 0.624, 0.285, and 0.031 for Intercept, Unemployment %, GDP growth %, Company ABC's Glassdoor rating, respectively.

Coefficients

From the Summary Output, the regression line is:

SUMMARY OUTPUT

Regression Statistics	
Multiple R	0.961
R Square	0.923
Adjusted R Square	0.890
Standard Error	1.203
Observations	11

ANOVA

	df	SS	MS	F	Significance F
Regression	3	121.498	40.499	27.963	0.000
Residual	7	10.138	1.448		
Total	10	131.636			

	Coefficients	Standard Error	t Stat	P-value	Lower 95%	Upper 95%	Lower 95.0%	Upper 95.0%
Intercept	13.708	9.814	1.397	0.205	-9.499	36.915	-9.499	36.915
Unemployment %	1.631	3.181	0.513	0.624	-5.890	9.152	-5.890	9.152
GDP growth %	1.316	1.136	1.159	0.285	-1.369	4.001	-1.369	4.001
Company ABC's Glassdoor rating	-2.795	1.036	-2.698	0.031	-5.246	-0.345	-5.246	-0.345

Based on the above coefficients,
- For each unit increase in Unemployment %, Company ABC's Attrition Rate increase by 1.631.
- For each unit increase in GDP growth %, Company ABC's Attrition Rate increase by 1.316.
- For each unit increase in Company ABC's Glassdoor rating, Company ABC's Attrition Rate increase by -2.795.

Regression formula

Based on the Summary Output, if "Unemployment %" is 3.3, "GDP growth %," is 2.3, "Company ABC's Glassdoor rating" is 5, then **predicted "Company ABC's Attrition Rate"**
= 13.708 + (1.631 * Unemployment %) + (1.316 * GDP growth %) - (2.795 * Company ABC's Glassdoor rating)
= 13.708 + (1.631 * 3.3) + (1.316 * 2.3) - (2.795 * 5)
= **8.1**

5.8 Word Clouds

To visualize the results of Sentiment Analysis, you can use graphs, histograms, and Word Cloud.

In this section, we will cover how to generate "word clouds" using the Microsoft Word. Word cloud (also called text cloud or tag cloud) is a visual representation of text data. Word clouds are more visually engaging than a table data - it is easy to understand and present, and is impactful as the most used keywords stands out graphically. Social media sites use word clouds to collect, analyze and share user sentiments. Marketers use word clouds to highlight the needs and pain points of customers. Data scientists use word clouds to report qualitative data.

Word Clouds is a popular way to visualize a message. In Word Clouds, the size of each word indicates its frequency and importance. Pro Word Cloud is a free Microsoft add-in to create Word Clouds.

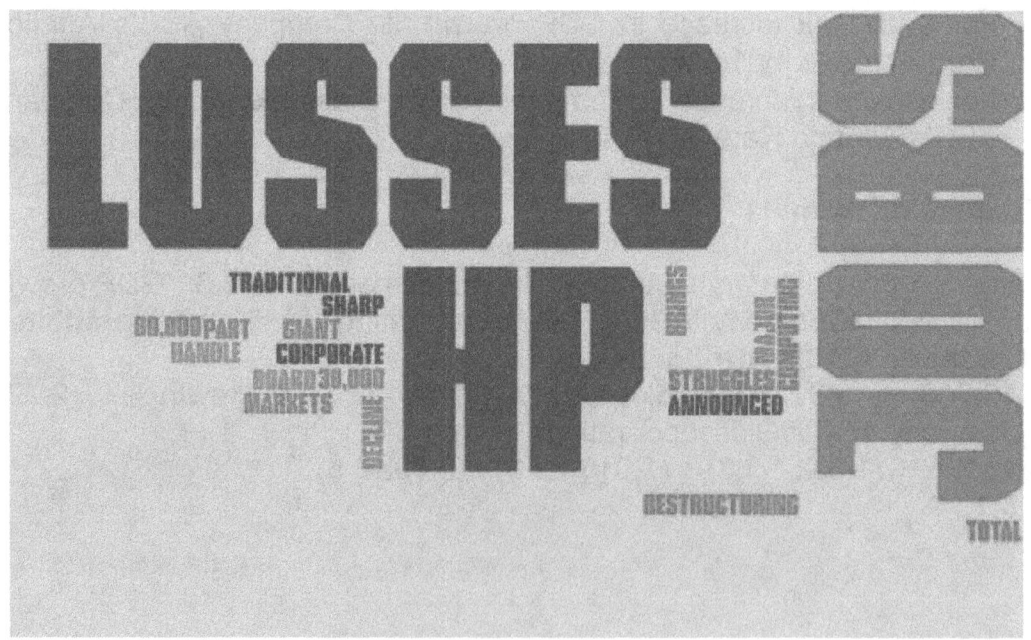

1) Copy and paste the text below, into your word document.

> Computing giant HP has announced a further 30,000 jobs losses as part of a major corporate restructuring. That brings to a total of 80,000 jobs losses as the HP board struggles to handle the sharp decline in its traditional markets.

Source: Jim Riley (2015), More Retrenchment at HP. https://www.tutor2u.net/business/blog/more-retrenchment-at-hp (8 March 2019)

2) Right click on a blank area at the top of your Word document, and choose "Customize the Ribbon".

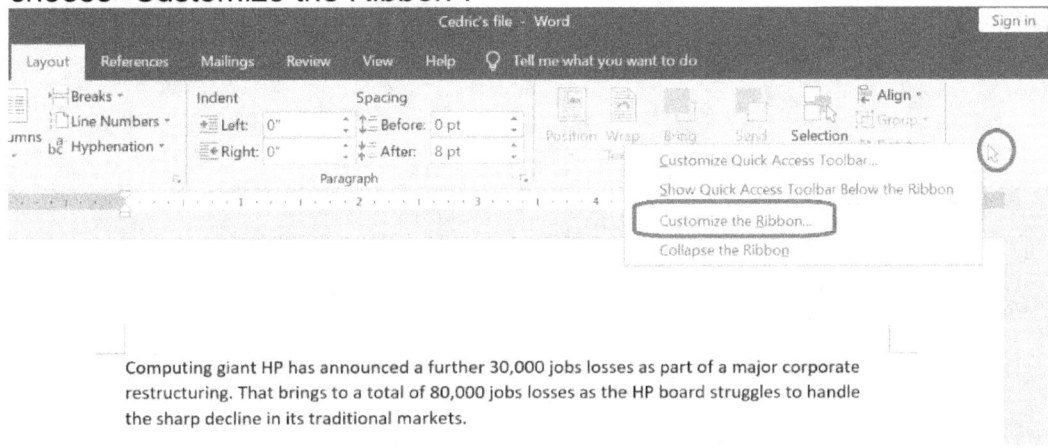

3) Tick "developer" and click ok, and you will see a Developer tab.

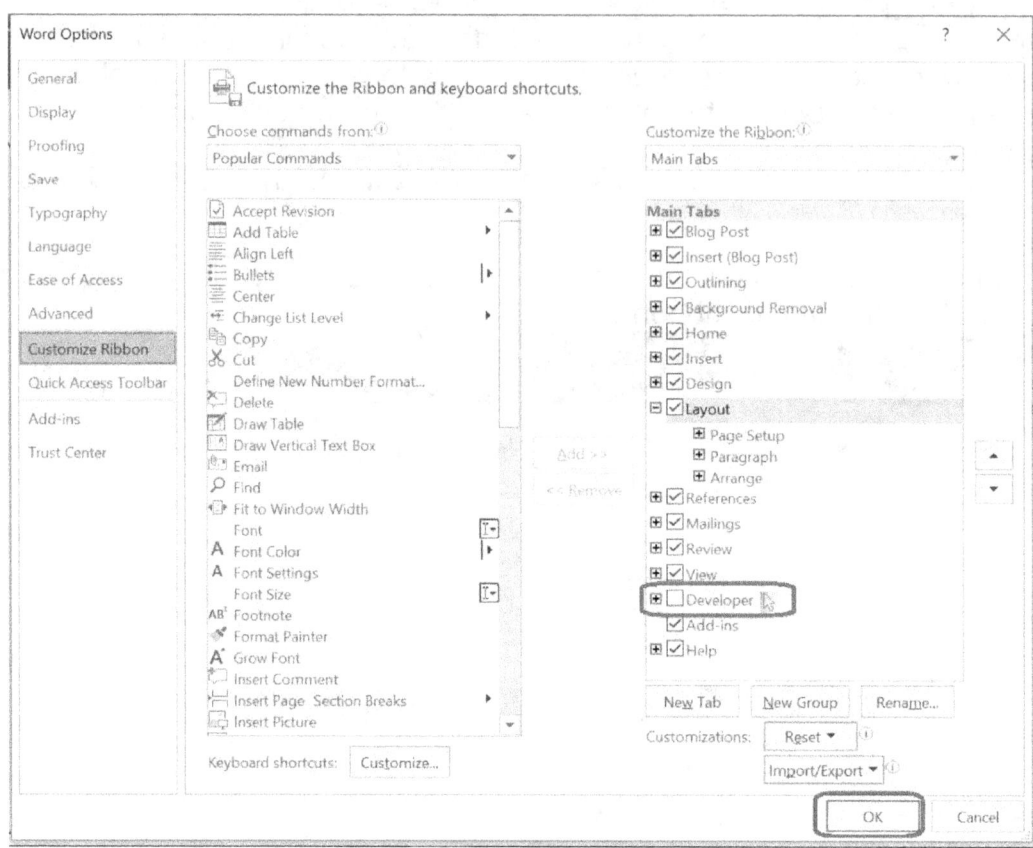

4) At the Developer tab, click the "Add-ins" button:

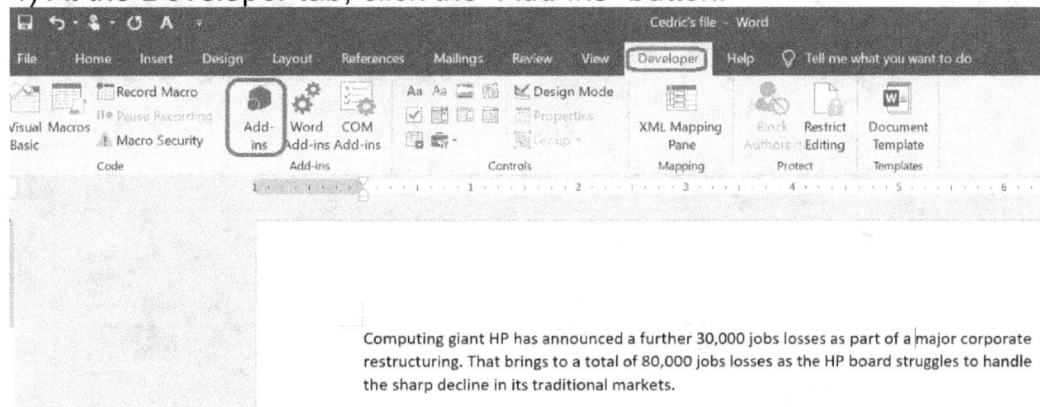

5) Click "Store":

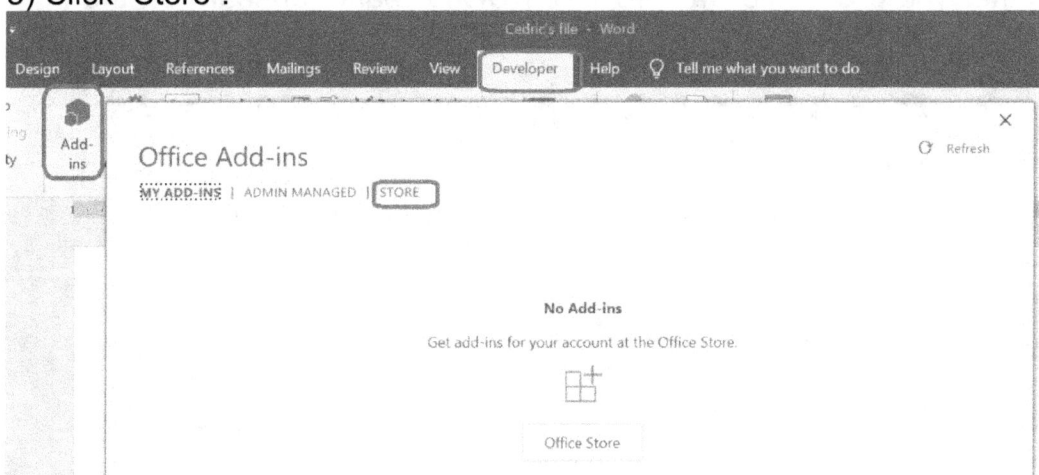

6) Find "pro word cloud", and click "Add".

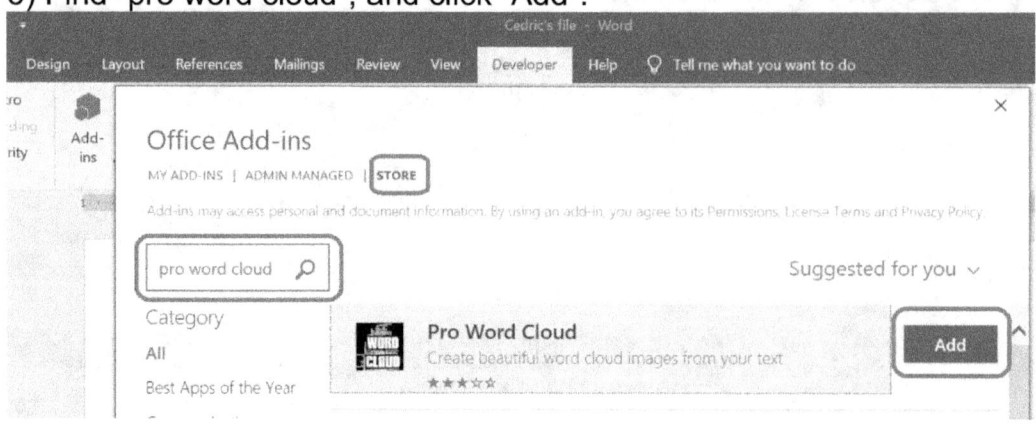

7) After you click "Add", a "Pro Word Cloud" panel appears on the right side of your Word document.

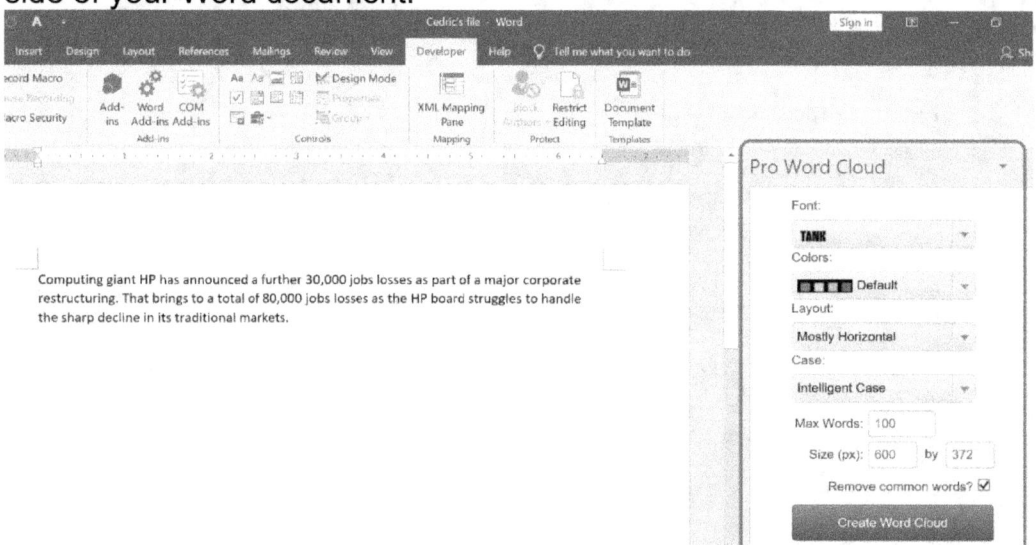

8) To create a Word Cloud, highlight the text that you want to turn into a Word Cloud, and click "Create Word Cloud".

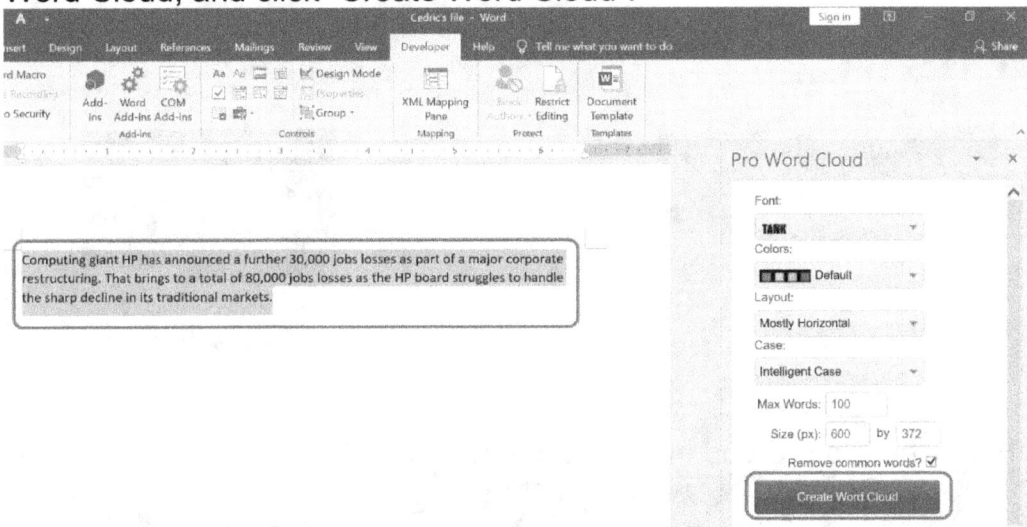

9) After you click "Create Word Cloud", a Word Cloud appears at the top right-hand corner of your Word document.

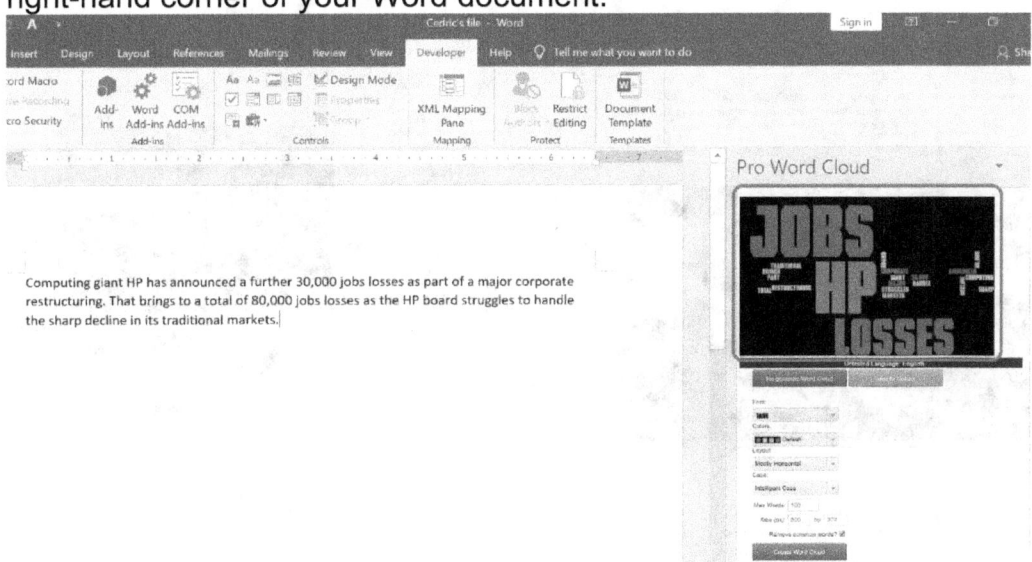

10) Right click on the Word Cloud, and click "Copy".

11) Paste the Word Cloud on your Word Document.

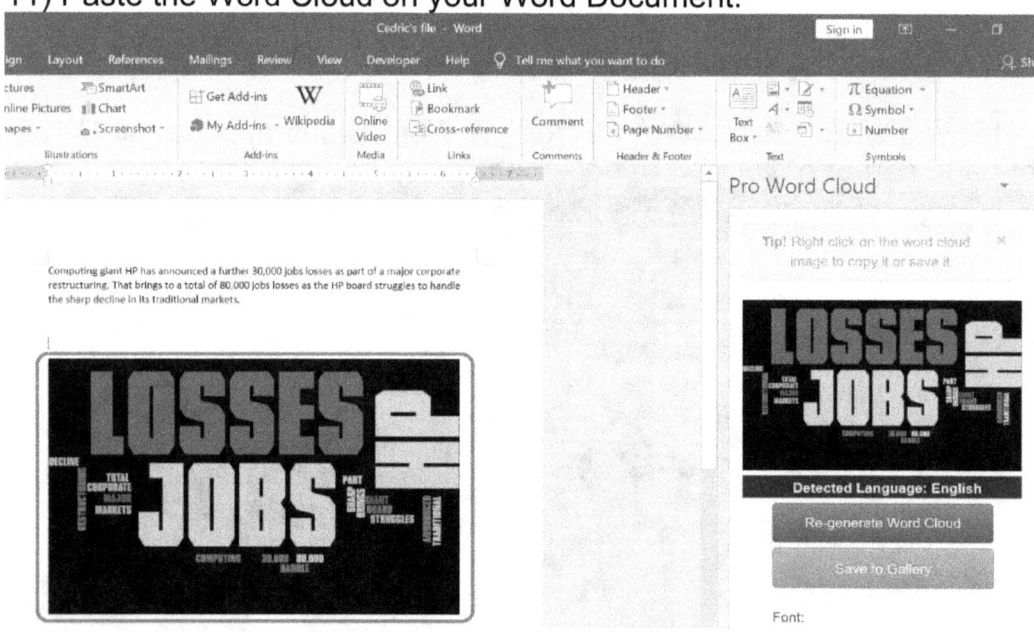

12) To format your Word Cloud, left click on your Word Cloud, click the "Format" tab, and you will see a list of formatting options.

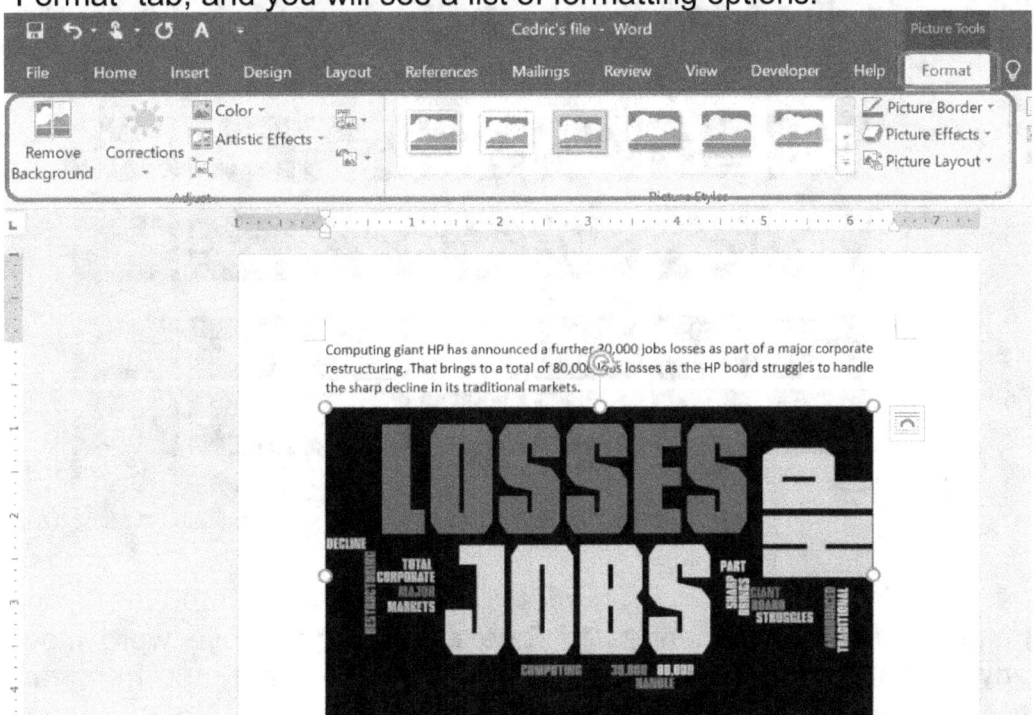

13) Click "Blue, Accent color 1 Light"

14) Your Word Cloud changes to Blue color. The above word cloud clearly shows that "HP", "Jobs", and "Losses" are the three most important words in the article.

5.9 IISS Engagement Prescriptions

The IISS model has a list of "Engagement Diagnosis Questions" and "Engagement Prescriptions" that corresponds to it's 5 Engagement Fertilizers (Basic Needs, Social Needs, Growth Needs, Meaning and Expectations). Engagement prescriptions are action plans to enhance employee engagement.

- **Engagement Diagnosis Questions** that you can use for sentiment gathering.
- **Engagement Prescriptions** that you can implement in your organization to build great employee experience & engagement.

IISS Model's Engagement Diagnosis Questions & Prescriptions

Engagement Fertilizers	IISS Engagement Diagnosis Questions	IISS Engagement Prescriptions
Fertilizer 1) Basic Needs	Q1) I believe I am paid fairly	Communicate pay philosophy, etc.
	Q2) I am satisfied with my company's employee benefits	Pantry full of snacks & fruits & drinks, dependent benefits, wellness benefits, etc.
	Q3) I am satisfied with the work life balance here	Allow work-from-home, provide flexible hours options, & ban after-work emails and messages, etc.
	Q4) The resources & processes support productivity	Streamline processes, availability of office stationery & meeting rooms with projector and telephone, etc.
	Q5) The physical working environment is appealing & conducive for work	Visually appealing physical working environment, quiet spaces for calls or just head-down work, etc.
Fertilizer 2) Social Needs	Q6) I work well with my manager	Include engagement skills in leadership development, etc.
	Q7) I work well with my colleagues	Help employees make friends at work. Facilitate social connection through technology. Hire for culture fit.
	Q8) I feel recognized for my work	Provide recognitions options (Thank you email templates, ePraise cards, Recognition App), share Recognitions ideas, etc.
Fertilizer 3) Growth Needs	Q9) I have opportunities to learn and grow	Provide various development modes. E.g. Gamification, Lunch & Learn, Mobile learning, Motivational talks, Orientation & Onboarding, Training, Mentoring, Job rotation, etc.
Fertilizer 4) Meaning	Q10) My job is meaningful	Job redesign
Fertilizer 5) Expectations	Q11) My job expectations are aligned with the organization	Set clear mutually agreed objectives, Acknowledge employee's struggles, Give Feedback and Ask for Feedback, Conduct stay interviews, Practice humility
	Q12) I will recommend this organization to my friends	Provide shareable social media content, etc.
	Q13) My manager sets clear goals aligned with company	Set clear mutually agreed objectives

https://www.amazon.com/dp/B0859CB567

By Cedric Ng ©

Annex 1) Publications by Author

https://www.amazon.com/s?k=ng+mong+shen
https://www.facebook.com/thehrdiary/

Index

A

absenteeism, 260
Absenteeism, 44
Actively disengaged, 36
Analytics Maturity Model, 53, 54
Aon Hewitt survey questions, 243
Attrition, 44
Autonomy, 124, 127, 130
Autonomy, Mastery, Purpose Framework, 126
Azure Machine Learning, 314

B

Bag 1, 15, 40
Bag 2, 15, 106
Bag 3, 28, 236
Bag 4, 30, 255
Bar Charts, 283
Basic Needs, 15, 22, 132
Belonging Needs, 111
Benefits, 139, 141
Best Buy, 66

C

Change management, 78
cNPS, 48
Coaching, 174
commute, 147, 148
Compensation, 133
Competence, 124
Connection, 130
Correlation, 58, 292, 325

D

Data storytelling, 76
Demographics', 45
descriptive analytics, 55
Descriptive Analytics, 54, 55
diagnostic analysis, 55
Diagnostic Analysis, 55
Diagnostic Analytics, 54
Diligence, 79

E

Employee Engagement, 34, 39
Employee Experience, 33, 34
Employee Life Cycle, 35
Employee Net Promoter Score, 37
Employee Satisfaction, 34
Encouragement, 80
Engaged, 36
Engagement, 42
Engagement & Motivation Theories, 108
Engagement Dashboards, 267
Engagement MAGIC model, 130
Engagement Metrics, 259
Engagement Metrics & Dashboards, 259
eNPS, 37, 48
Equity Theory, 225
Esteem Needs, 111
Expectancy Theory, 223
Expectation Gap, 228
Expectations, 15, 16, 26, 221
Extrinsic Motivations, 127

F

Fertilizer 1, 22, 132
Fertilizer 2, 23, 150
Fertilizer 3, 24, 162
Fertilizer 4, 25, 189
Fertilizer 5, 26, 221
Flexible work options, 144
Focus groups, 252
Frederick Herzberg's Motivation Theory, 117

G

Gallup, 36, 44
Gallup survey questions, 241
gig contractors, 143
Glassdoor, 253, 325
Gossip, 83
grievances, 260
GROW Model, 174
Growth, 130

Growth Needs, 15, 24, 162

H

Health, 45
Help people make friends with others, 154
Help people make friends with you, 158
Heroes, 91
humility, 234
Hygiene factors, 118

I

IISS Engagement Diagnosis Questions, 250
IISS Engagement Prescriptions, 348
Impact, 130
Innovation, 45
Inspire Engagement Investment, 18, 40
Inspire with Engagement Fertilizers, 15, 20, 106
Inspire with Engagement Investment, 15
Intrinsic Motivations, 127
Inventory, 45
ISS, 46

J

Job Characteristics Model, 198
Job rotation, 171
Job security, 147, 148

M

Maslow's hierarchy of needs, 110
Mastery, 128
Meaning, 15, 25, 130, 189
Meaning in employee value proposition, 215
Meaning in engaging work, 197
Meaning in organizational values, 207
Meaning in work that fits with your life, 206
Meaning in work that helps others, 200
Meaning in work you're good at, 202
Meaning with friends at work, 204
Meaningful work, 195
Mentoring, 175
Mobile learning, 168
Motivating core-performers, 135
Motivating high-performers, 137
Motivating under-performers, 136

Motivation factors, 118
Motivational talks, 169

N

Narratives, 76
Net Promoter Score, 37, 259
Not-engaged, 36

O

objectives, 231
Officevibe survey questions, 246
onboarding, 169
Organisational engagement survey, 239
Organizational Climate, 39
Organizational Culture, 39
Orientation, 169

P

participation rates, 266
passion, 194
Patience, 84
Perception of pay, 134
Physical working environment, 147, 148
Physiological Needs, 111
Predictive Analysis, 56
Predictive analytics, 56
Predictive Analytics, 54
Prescriptive Analysis, 56
Prescriptive analytics, 56
Prescriptive Analytics, 54
Prizes, 135
processes, 147, 149
productivity, 266
Profitability, 46
Pulse surveys, 251
Purpose, 128

Q

Quality, 47
Quality of friends, 152

R

Radar Charts, 287

Reasonable working hours, 147, 149
Recognition, 160
Regression, 66, 300, 332
Relatedness, 124
Risk taking, 86

S

Safety, 47
Safety Needs, 111
Sales, 47
Salespeople, 135
Self-determination Theory, 123, 130
Sentiment analysis, 309
Sentiment Analysis, 16, 314
Sentiment Diagnosis & Prescription, 16, 30, 255
Sentiment Gathering, 16, 28, 236
Sentiment Word Cloud, 312
Service, 48
Social media, 339
social needs, 151
Social Needs, 15, 23, 150
stay interviews, 232
stories, 90
Stress, 193

T

Teach others, 165
Teamwork, 87
text cloud, 339
Total Shareholder Returns, 49
Train Managers to Lead, 166
turnover rate, 261

V

Victims, 91
Villains, 91
visuals, 94
Visuals, 76

W

Wellness, 145
Word cloud, 339
Work-life balance, 89

www.ingramcontent.com/pod-product-compliance
Lightning Source LLC
Chambersburg PA
CBHW060409220526
45465CB00008B/2822